YOUR DAY, YOUR WAY

The Fact and Fiction Behind Your Daily Decisions

TIMOTHY CAULFIELD

RUNNING PRESS
PHILADELPHIA

Running Press
Hachette Book Group
1290 Avenue of the Americas, New York, NY 10104
www.runningpress.com
@Running_Press

Printed in Canada

Originally published as *Relax, Dammit!* in 2020
by Penguin Canada in Canada
First US Edition: December 2020

Published by Running Press, an imprint of Perseus Books, LLC, a subsidiary of Hachette Book Group, Inc. The Running Press name and logo is a trademark of the Hachette Book Group.

Print book cover by Amanda Richmond.

Library of Congress Control Number: 2020942770

ISBNs: 978-0-7624-7249-9 (paperback), 978-0-7624-7248-2 (ebook), 978-1-5491-5938-1 (audio book)

LSC-C

1 2020

To science. Hang in there.

YOUR DAY, YOUR WAY

Also by Timothy Caulfield

The Cure for Everything!: Untangling the Twisted Messages About Health, Fitness, and Happiness

Is Gwyneth Paltrow Wrong About Everything?: When Celebrity Culture and Science Clash
(also published as *The Science of Celebrity . . . or Is Gwyneth Paltrow Wrong About Everything?)*

The Vaccination Picture

CONTENTS

PART III: EVENING

INTRODUCTION

DECISIONS, DECISIONS . . .

"You're going to feel pretty bad if your son dies! You're going to feel horrible!"

This was my sister-in-law's mic drop retort to an intense debate we were having about risk. It happened at a family Sunday dinner. I had just told the collected kinfolk that I had decided to go skydiving. Ten thousand feet, mostly free fall.

Everyone thought I was crazy. I am pretty sure most of them entertain this sentiment on a fairly regular basis, so the thought of Uncle Tim plummeting to the earth at terminal velocity wasn't what ignited the debate. What really upped the family angst was that I planned to take my 14-year-old son, Michael.

We make a ridiculous number of decisions every day. Some estimates are that the number hovers in the thousands; we make hundreds of decisions daily about food alone. We make decisions about when to wake up, how to brush our teeth, what to have for breakfast, the amount of coffee to drink, how to get our kids to school, and on and on.

This book is structured around the decisions we make throughout a typical day, from when we wake up to when we go to sleep. In between, I explore dozens of different choices. Some touch on issues that are relatively frivolous and fun: Should you sit on a public toilet seat? What is the best way to park your car? Is ranting to your work colleague a good idea? Should you cuddle after sex? Others are more serious and controversial: Should you let your kids walk to school? Should you step on a scale in the morning? Should you feel guilty about not spending enough time with your kids? For some of the topics I provide an overview of the history of a particular decision and an analysis of the social forces distorting the evidence and public discourse. Others I deal with briefly, delivering just the facts. The goal is to provide a useful summary of what the evidence says about a particular decision, but also to give insight into how cultural, historical, and scientific forces take hold of and shape our thinking on a wide range of issues relevant to our day-to-day lives.

I realize that relying on evidence isn't the only way people make decisions, but I think that a journey through a typical day of decisions will allow us to recognize that, more often than not, we can relax. It probably doesn't really matter as much as we think. In a world where information is increasingly twisted for commercial, ideological, and personal gain, finding a path to the objective truth on any topic, from toothpaste to toilet seats, can be difficult. But the path does exist, and finding it can be liberating.

But first, let's look at what goes into making our decisions.

Making decisions is tough work. Indeed, it tires us out. It can cause stock analysts to perform progressively worse over the course of a day. It can lead us to make poor choices about what to eat (the more tired our brains are, the more junk food we consume). It can change how physicians prescribe drugs and how judges handle sentencing. And the more deliberate the decisions—that is, the more we think through them—the more fatiguing they are.

Decision-making is a complex, messy activity that can lead to significant stress. But it doesn't have to. One of the goals of this book is to remind us not to fall prey to the numerous social forces that increasingly turn making a decision into an unnecessarily anxious process. If we can look past the popular culture noise, marketing pressures, and ideologically motivated spin, we can often find a science-informed, and less stressful, way forward.

And I hope to also provide useful context about our daily choices, allowing you to look at many of them in a new light—even if it doesn't change your mind. But what this book is really about is the justifications behind our decisions and the cultural, historical, and scientific forces that shape the evidence we use to inform them.

We all want to make the *right* decisions—or, at least, the decisions that are right for us. We select a particular food because we believe it is healthier or better for the environment. We brush and floss our teeth because we don't want them to fall out. We say no to another cup of coffee because too much caffeine might be bad for us. We drive our kids to school because we worry about their safety. We avoid sitting on public toilet seats so we won't pick up germs. We take a nap in the afternoon because we've been told it will boost our productivity. We try not to obsessively check our email, as we've been told doing so will stress us out. We stand because sitting is the new smoking. We take vitamins because we want to stave off disease. We vent our rage because we have been told a cathartic release is psychologically beneficial. And if we are asked about why we made a particular decision, most of us will muster some relatively coherent rationale: it is healthier, safer, tastier, or, simply, better.

But as we will see, many of these rationales and beliefs do not fit with the evidence. From the moment we wake to the moment we drift off to sleep at night, we make dozens and dozens of decisions that are based, to a lesser or greater extent, on misinformation.

I am not saying that we are all hyperrational beings who seek to only make decisions that accord with the facts. On the contrary, innumerable cultural, social, and psychological forces shape the decisions we make. And our rationales for a decision are often tacked onto our actions *after* we have made our decision. That is, we make a decision first and then, consciously or unconsciously, we construct reasons why it was the correct one, reasons that fit the decision, our personal identity, or our desire to be seen as consistent or logical—even if the original motive for the decision might be different or even unknown to us.

We do this as a way to avoid what is called post-decision dissonance. Studies have found that after choosing, for example, a new job or which university to attend, people will rank the attributes of their chosen institution higher than they did before making their decision and—no surprise—rank the attributes of the loser lower. In addition, we make decisions that allow us to feel we are being consistent. So past decisions shape future decisions. For instance, perhaps you bought GMO-free food because you already buy organic food and that seemed like a consistent pattern and fit with your personal image. You then need a rationale to justify this decision, even though what influenced your food choice was simply the drive to be consistent. And once the decision is made, you must become more committed to the rationale, because now it is part of how you perceive yourself.

We *all* do this kind of thing. You do it. I do it.

Having perfect knowledge of the available science will not lead to a world of strictly science-informed decisions. Still, exploring the deep disconnect between the rationales and beliefs behind our decisions and what the actual science says provides an opportunity to dive deep into three harmful, and increasingly pervasive, social paradoxes. Indeed, I believe these paradoxes have become significant and less than laudable hallmarks of our time. And all three make decision-making more difficult and infinitely more stressful.

First, there is the *knowledge-era paradox*. We live in a world awash in information. There has never been as much science-informed knowledge as there is now. It has been estimated that the global scientific output doubles every nine years. Since 1665, when the first academic journals were produced in France and Britain, more than 50 million scholarly articles have been published. And approximately 2.5 million new scientific papers are published each year. Not all of these papers are useful to our day-to-day decision-making, of course, but these numbers give a sense of the growth and quantity of scientific inquiry. There is a lot of information out there. And there has never been such easy access to that information. So you'd think that our decisions would increasingly be informed and evidence-based.

But there has also been a growth in the number and sway of social forces perverting that knowledge. This is the era of fake news, conspiracy theories, alternative facts, and social media confirmation bubbles. Too often information is twisted, hyped, misrepresented, or overtly misinterpreted. So, in fact, more and better-available information often does not lead to better-informed decisions. Looking at what the evidence says or doesn't say about a particular decision allows an exploration of how and why information is skewed. And this analysis will, I hope, help you tease out what to pay attention to and what to ignore.

Second, there is the *less-risk paradox*. A lot of the decisions we make are to some degree about avoiding harm or minimizing risk. No matter how seemingly innocuous the choice—when we go to bed, what we eat, how we get to work, even how we wash our hands—the concept of risk often plays a role. And advertisers know this. Fearmongering has become a dominant theme in the marketing of products and ideas, particularly in the context of health. The multibillion-dollar wellness industry creates a reason to be fearful—such as the need to avoid toxins—and then markets a solution to that manufactured fear.

Entire industries are dedicated to the goal of keeping our kids safe. There are monitoring devices aimed at stopping stranger-danger abductions and merchandise to reduce your child's exposure to the myriad chemical substances that are, or so the marketing tells us, threatening their health. Jessica Alba's The Honest Co., valued at over $1 billion, appears to have been founded on the philosophy that fear (plus parental guilt) sells: the company markets things like diapers (made with naturally derived materials, of course), vitamins (organic and non-GMO, clearly), and cleaning products (hypoallergenic, obviously) based on its stated philosophy of providing "peace of mind" because "protecting you and the ones you love" is the company's top priority.

Yes, we need to be vigilant about potentially dangerous substances in our environment. But we are not living in some dystopian Mad Max world awash in toxic chemicals that will melt our skin and mutate our DNA. In fact, in most countries, this is the safest and healthiest time to be alive, even if we're not convinced of that. By almost any objective measure, things are better now than ever before. In the U.S., for example, both violent crime and property crime rates have fallen steadily and significantly over the past quarter century. Yet in 2017 almost 70 percent of Americans thought there was more crime than a year before. Only 19 percent got it correct: there is less crime.

Teenage pregnancy has declined. There are fewer missing children. Illicit drug use by teenagers has declined, too. Binge drinking and cigarette smoking are far less common than a few decades ago. And fewer child pedestrians are being hit by cars, even though there are more people, and more people driving more miles, than ever before. (In 2017, Americans drove a record 3.22 trillion miles.) It is now safer for your child to cross the street than when you were a kid yourself. Data from the U.S. National Highway Traffic Safety Administration shows that since 1993 the number of children being struck and killed by cars each year has declined from over 800 to less than 250.

It is the same with many diseases, too. Fewer people are dying of infectious diseases, thanks largely to vaccination—an invention that saves millions of lives every year. Even with cancer we often overestimate the actual risks and place blame in the wrong places (Wi-Fi, stress, GMOs). Although cancer remains a leading cause of death, second only to heart disease, here too things are improving. In 2018, the U.S. National Cancer Institute reported that between 1990 and 2014 the overall cancer death rate had declined by 25 percent.

We see a similar trend at the global level. Most people around the world believe that rates of poverty and child mortality are increasing, whereas both have decreased significantly over the past decades, and the improvements are accelerating. Since 1960, child deaths have declined from 20 million a year to 6 million a year. Globally, life expectancy is up 5.5 years since 2000. It is important to recognize progress. Yet few of us do. A 2015 survey found that only 6 percent of Americans think the world is getting better. The bottom line is that we live in an era of unprecedentedly low levels of health and safety risks, and yet we are, paradoxically, increasingly preoccupied with the elimination of risk, both real and imagined.

We should, of course, continually strive to improve things, such as the livability of our neighborhoods, the safety of our streets, access to healthy food choices, and the cleanliness of our air and water. And we are facing tremendous challenges, including climate change, eroding biodiversity, climbing obesity rates, mental health concerns, antimicrobial resistance, continued global conflicts, and less than ideal levels of social tolerance and diversity, just to name a few. So I'm not arguing that things are perfect—only that we need to keep things in perspective. Again, few people do.

There are many reasons why we believe risks are everywhere—including the fact that we are hardwired to spot and want to avoid risks—but, as I will argue, we should not let fear dominate our decision making. Indeed, there are reasons to believe that our obsession with risk and our health is actually making us *less* healthy and happy. Scientists have even found an association between anxiety about health and a substantially increased risk of heart disease.

Many of the decisions we make to lower risks to our health and safety in fact have the opposite effect. For me, this is one of the most perplexing contradictions of our time. There are absurd examples, like IV vitamin infusions and colonics. Those are science-free and dangerous practices, often touted by practitioners of alternative medicine and celebrity health gurus as a way to detoxify our bodies and improve our health. But there are many other more common practices—including decisions about our sleep, what we eat, and how we transport our families and ourselves—that are not achieving the risk-reducing and safety-enhancing results we think they do.

Finally, the third paradox that makes decision-making more difficult and more stressful is the *perfection paradox*. We are under increasing pressure to improve ourselves and to continually strive toward perfection—in our appearance, careers, relationships, parenting, mental health, sex life, you name it. This compulsion has become the modern, perverted version of Thomas Jefferson's rights to life, liberty, and the pursuit of happiness. But it is a compulsion viewed not only as a right but also as a social responsibility. Recent research, including a large 2017 analysis of more than 40,000 college students, has found that both perfectionist attitudes and perfectionist social pressures have increased markedly over the past decades. The paradox is that, in general, neither the journey toward perfection nor the attainment of a perfection-oriented goal confers greater happiness, at least in the long term. Rather, it is associated with a host of physical and mental health issues.

Many of the goals we are nudged to pursue, whether they are about our physical appearance, our relationships, or our careers, are entirely illusory. People are striving for a perfection that does not exist and, as such, is unattainable. To cite just one example, new social media apps allow anyone to alter and perfect how they look online. This has resulted in completely unrealistic standards of beauty. Celebrities and models have long been airbrushed and photoshopped. Now everyone can tweak their image. To make matters worse, social media and mobile devices create the opportunity for near-constant social comparison, something that leaves many of us miserable, anxious, and dissatisfied.

Why is this perfection paradox relevant to a book about decisions? We make a host of choices every day—sometimes expensive and time-consuming ones—meant to bring us closer to some ideal. Some of the increasing social pressures to self-improve come from the entities marketing products to help us in that quest. Not only are these perfection-seeking decisions often pointless, but the associated choices—about, say, what supplement to take or self-help strategy to employ—are not informed by science. When this happens, it is a triple-whammy waste of your time, energy, and (often) money.

My wife Joanne was not pleased. "I don't mind if you go, but do you need to take Michael?"

Everyone I told about my skydiving experiment agreed with Joanne. Absolutely everyone. The consensus was that I was crazy to involve my son, that this was a very bad and irresponsible parenting decision. Michael, however, couldn't wait. He located the best local skydiving outfit to provide the death-defying service and nagged me constantly to make a reservation. "The weather looks pretty good this weekend, Dad. We better book a spot," he'd say several times a day.

I'm not saying skydiving is a *good* idea. In fact, it terrifies me. I was absolutely horrified by the idea of throwing myself out of a plane—let alone my youngest son jumping alongside me. I hate flying in small planes; they tend to make me vomit. And I am not, by any measure, a particularly brave individual. I'm the kind of person who at the sight of a black bear bolts and leaves his beautiful young bride to be eaten. In fact that is precisely what I did when, early in our marriage, my wife and I came across a black bear. But these are the exact reasons I decided to go skydiving. Despite my genuine fear, I knew it was a relatively safe activity. Skydiving is the perfect example of a scary activity that isn't nearly as dangerous as we think. How safe is it? According to the United States Parachute Association, there are an estimated three million jumps each year. The fatality count, per year, hovers around 20. That is a 0.007 percent chance of a death for every 1,000 jumps. That is a very, very small risk. It was way more dangerous driving to the jump site (or for that matter to the corner store). With tandem jumping—which is when you are strapped to an expert and what beginners like Michael and I would do—the fatality rate is even lower. According to the British Parachute Association, "The all-time tandem fatality rate since 1988 is about 0.14 per 100,000 jumps." Put another way: 1 in 703,000. This is a very, very, *very* small risk. And in the U.K., there have been zero tandem fatalities in the past 20 years. The injury rate for skydiving, which is defined as everything from a minor scratch to fractures and sprains, is also remarkably low, at about one in 1,100 jumps.

Michael is a competitive gymnast. He's at the level where he does swings on the top of the high bar, iron crosses, twisting dismounts off the rings, and spinning flips on the floor and vault. His chance of a serious injury in this sport, if you believe the research, is more than 10 times greater than getting even a minor scratch skydiving. But none of the people who scolded me about my skydiving decision have ever warned me off gymnastics. Some of the scolders have kids who play hockey, an activity that seems to me more dangerous than either skydiving or gymnastics: one study found there are approximately 19 injuries (defined as fairly serious things like concussions and ligament damage) per 1,000 games played.

Again, the point is not that gymnastics or hockey or driving are bad decisions as compared to jumping out of a plane, but it does show how we all weigh and think about risks in weird and complex ways, especially when our kids are involved.

And by the way, being killed by a bear is also a pretty remote danger—so my wife-sacrificing, dignity-destroying fear is also irrational. Between 1900 and 2009 there were 63 fatal bear attacks in North America. If you think of the millions and millions of individuals exposed to bear habitats over that period, it gives you a sense of how remote the risk actually is.

But when a bear attack does occur, it is almost always big news that generates headlines, TV coverage, and follow-up personal interest stories on the attacked people and the community. So you perceive the risk as greater, and when you are walking in the woods with your romantic partner, it is easy to imagine the gruesome event happening to you.

This kind of thinking is a big reason why we focus too much on remote health and safety threats. When an event is easily recalled or produces a particularly strong emotional response, we tend to emphasize the risks associated with it. It is easier to picture a gruesome bear mauling or a skydiving accident than, say, the well-documented ill effects of not eating enough fruits and vegetables or not getting enough sleep. So we avoid the bear but don't worry, moment to moment, about not consuming enough carrots or missing sleep. These are well-known cognitive biases called the availability heuristic and the affect heuristic. These hardwired shortcuts to how we think about the world are just a few of the reasons why we all make less than logical decisions about how to minimize risks. And this is why skydiving strikes many as a bad idea. It is an extreme activity that can easily conjure thoughts of a horrid outcome. Splat.

Unfortunately we live in a media world that is, as research has consistently demonstrated, a risk amplifier and anxiety generator. And things are getting worse. A 2018 study from the University of Warwick found that stories about risk—such as disease outbreaks, terrorism, or natural disasters—become increasingly negative, inaccurate, and hyperbolic as they are shared. It is like a perverse and perverting game of telephone. As the senior author of that study, Professor Thomas Hills, told me, "When people hear information, they tend to selectively reproduce the scary bits. Because everyone does this, it precipitates the information down to what scares us, amplifying the apparent risks."

Professor Hills, whose research involves exploring how information is used in decision-making, told me that today's world has been called the risk society "because it seems all we care about anymore is how to minimize risks . . . This makes risk *the reason* for doing or not doing anything, and impairs objective assessments of pros and cons."

While risk is clearly a big part of the story, many other social forces affect the information we use in our decision-making. Recent research on the dissemination of information on social media has found that falsehoods "diffused significantly farther, faster, deeper, and more broadly than the truth"—likely because lies are simply more interesting than the truth. In total, the combination of cognitive bias and the spread of twisted information makes it near impossible to tease out what the best, most evidence-informed decision actually is.

Given our current media environment, it is easy to get caught up in the whirlwind of misinformation. It is easy to let misinterpreted risk, hyped science, conspiracy theories, and health myths cloud our decision-making. The good news is that for many of the decisions we make hour to hour and day to day, we can turn to science. If we know what to look for, we can make evidence-informed choices.

The airplane was tiny and old, a Cessna from the 1950s. It was also extremely loud. The seats had been removed, and Michael and I were jammed into a space right behind the pilot. It was so cramped I couldn't sit up or straighten my legs. Waves of claustrophobia simmered in the back of my skull.

The plane groaned and shuddered as it climbed to the prescribed altitude. It felt like we were in a flying lawn mower. It felt like this had been a very bad decision.

When we got to 12,000 feet, the engine slowed and the pilot flung the door open. There was a roar of wind and even more noise. Michael and I locked eyes. "Holy shit," he reflexively mouthed, a huge smile on his face.

Michael and the instructor inched their way to the door. Michael put his feet out of the plane and before I could process what was happening, he was gone. My stomach churned as I saw my son fall, in what seemed like an instant, to the size of a small dot.

What the hell had I been thinking?

Then it was my turn. Without thinking—because that would result in a cowardly retreat—I placed my feet on a little platform outside the plane. The wind was intense. I found it hard to breathe.

We tumbled out of the plane.

"Ahhhhhhh!" I screamed uncontrollably. "Amazing! This is amazing!"

I

MORNING

6:30 AM—Wake up!

Unless you are one of those irritatingly peppy morning people, dragging yourself out of bed is often one of the hardest things you will do on any given day. But when, exactly, *should* we wake? Does the early bird really get the worm?

My father couldn't stand the thought of his kids sleeping in. Even if there was no reason for me to get up—school, work, chores—he would burst into my room early in the morning and arhythmically tap his finger on my shoulder, declaring I was "wasting the best part of the day" or that I should "get up and do something useful" (or some other parental-advice cliché). Which only added aggravation to this morning ritual. To this day, I cringe when someone pokes me in that particular arhythmic manner.

Is there any evidence to support my father's get-the-hell-up commandment?

"For me, when it comes to waking up, consistency is the most important thing," Professor Satchidananda Panda told me. "In fact, your day really starts the night before with when you go to bed."

Professor Panda is a researcher at the Salk Institute for Biological Studies in California and a renowned expert on the human circadian rhythm, which is the master biological clock that helps to regulate how our bodies function. He has studied the influence of these daily rhythms in cells, flies, mice, and humans. "We should try to go to bed as close to the same time every night," he advised. "And we should also wake up around the same time."

It was at night that he passed along this sleep advice to me. He had a glass of wine in his hand and a smile on his face, so there was a tinge of irony to the recommendation. We were both speaking at an academic conference in Santa Barbara and, at that moment, we were enjoying some post-event downtime. Earlier in the day Professor Panda had presented his compelling research on how ignoring our circadian clock can have a profound effect on our well-being. "Of course, being consistent can be very difficult," he continued, raising his glass.

Numerous studies back up Professor Panda's assessment of the value of a consistent sleep pattern. Much of this research has been done on university students, a cohort not known for adherence to a strict sleep schedule. One 2017 study of undergraduate students found that those with irregular sleep patterns had lower academic performance. Another 2017 study of more than 100 undergrads found that sleep regularity was associated with increased general well-being. Not surprisingly, being consistent with your turning-in schedule is also associated with sleeping longer and better, and not just for chronically exhausted students. A 2018 Australian study of over 300 elderly individuals—a demographic that often struggles to sleep well—determined that sleep regularity can help people get the recommended amount of sleep, which, for most adults, is around seven hours a night (though there is significant variation). An irregular sleep schedule has also been linked with poor dietary choices and weight gain.

So, was my father right? Was his exasperating finger-poking wake-up strategy doing me good by keeping me on a consistent sleep schedule?

Not exactly.

Studies have found that morningness is correlated with such things as greater life satisfaction, academic performance, and workplace productivity. Indeed, waking early is often seen as a successful approach to life. It is what winners do. It is what hard-driving executives do. Apple CEO Tim Cook reportedly wakes at 3:45 AM. When she was first lady, Michelle Obama woke at 4:30 AM. Quarterback Tom Brady is watching football film at 5:30 AM. And actor Dwayne "the Rock" Johnson is in the gym, lifting ridiculous amounts of weight, at 4 AM.

But from a health and productivity perspective, this is not, contrary to the pop culture push, a good way to organize your life—at least for a significant portion of the population. There is a remarkable disconnect between the accepted wisdom and what the research tells us about when to wake. Some researchers have even suggested that our entire society has set the work clock wrong. Dr. Paul Kelley, a researcher in sleep, circadian, and memory neuroscience at Open University in the U.K., argues that we should reconfigure when we start work and school, because our bodies have a natural biological rhythm that does not fit our current timetable. Given that some of us are morning people and some are night owls, Dr. Kelley believes that for a large hunk of the population, starting work at 10 AM, which is pretty darn late for many of us, would actually *increase* productivity. Getting up too early—a near-universal norm, Kelley argues—messes with our brain in a manner that reduces creativity and increases the risk of mistakes. It may also contribute to mood problems, weight gain, and a decrease in life satisfaction. In some jobs, it can create work site safety issues.

"I'm actually an extreme morning person," Dr. Kelley told me. "I love the morning. The light. The quiet. And I'm a train wreck by 8 PM. Yes, I am fully aware of the paradox. I am advocating for social change, despite the fact that I benefit from the current situation. But given all the evidence, it is the right thing to do."

The fact that morning people, including Dr. Kelley, *seem* more successful may simply be the result of our society's being structured to benefit morningness. The alleged advantages of being an early bird are deeply ingrained. Rising early is almost always portrayed as a virtuous and noble act. If, on the other hand, you wake later than what is generally considered a reasonable time, you are viewed as lazy or unmotivated. (Studies have put the average wake time in the U.S. at 6:37 AM, though there is considerable country variation, with Slovenia waking at 6:02, Canada at 6:50, and Argentina at 8:44.) And think of all the famous quotes and aphorisms espousing the value of waking early. Thomas Jefferson said, "The sun has not caught me in bed in 50 years." Aristotle advised, "It is well to be up before daybreak, for such habits contribute to health, wealth, and wisdom." And Benjamin Franklin's singsongy version of that same prescription is perhaps the most famous of all: "Early to bed and early to rise, makes a man healthy, wealthy, and wise."

Can you think of a single saying promoting the idea of staying in bed a wee bit longer? Perhaps we should make up a few. How about, "Those who sleep in are destined to win!" Or, "Midmorning waking and a productive day you will be creating!" Maybe, "Ignore the clock and finish on top!"

During our discussion, Dr. Kelley suggested that in most of the world there is a systemic bias against eveningness. Some people are, through no fault of their own, at a distinct disadvantage because they aren't wired to embrace the morning. "There has long been an assumption that everyone is good with an early start time," he said, clearly frustrated by the situation. "This is simply not true! This bias can have real negative consequences."

Your propensity to rise early or not—that is, your chronotype—is largely determined by genetics. A 2017 analysis of genetic data, including an analysis of twins, concluded that genetic factors explain up to 50 percent of your chronotype. Brain scans have also found structural differences in brains associated with different chronotypes. So it is safe to conclude that much of when you wake up comes down to born-with-it biology. This means changing it significantly—that is, forcing yourself to become a lover of the early AM—isn't easy or even advisable. My friend and colleague Dr. Charles Samuels, the medical director of the Centre for Sleep and Human Performance at the University of Calgary, told me that it is possible to shift your chronotype slightly, perhaps an hour or so, if strategies are applied consistently—and the core strategy is, again, being consistent.

Although it isn't a character flaw to not be a morning person, a 2018 study of more than 400,000 people did find that being a night owl "was associated with a small increased risk of all-cause mortality." But, again, that may be because the schedules of modern societies are structured around the morning-works-for-everyone assumption. A 2018 study explored the relative contributions of hardwired biology and lifestyle factors to the health risks associated with being a late-night person. The researchers concluded that the health risks were caused, not by the biology that creates the propensity to stay up late, but by the behaviors associated with those late nights (such as late-night eating, poor sleep, and being sedentary). This is good news, as these are modifiable behaviors, especially if accompanied by social changes that allow for a bit of flexibility to accommodate different chronotypes.

Why and how chronotypes affect our health are a complex puzzle that researchers are still trying to figure out. For the purpose of our day-to-day decisions, the important point is that there is now a broad consensus that chronotype does matter to our health and well-being.

There is even more evidence to suggest that the biological clocks of teenagers are ill suited to the current daily schedule. A 2017 study followed over 30,000 students for two years and found that starting school later was associated with an increase in grades, more successful graduations, and, not surprisingly, better overall attendance. Dr. Kelley points to evidence that indicates that later start times could increase academic performance by as much as 10 percent. And allowing teenagers to sleep more can have other social benefits. Brain imaging research from 2013 concluded that when teenagers don't sleep enough—which is the case for approximately 80 percent of them—their inhibitions are reduced even more than usual for teens, resulting in greater, and possibly dangerous, risk-taking behavior. This is why many scholars, including Professor Kelley, have been pushing for later start times for school. And as the evidence mounts to support his position, we are seeing jurisdictions throughout the world take this issue more seriously. A 2017 study from the RAND Corporation concluded that using later school start times would contribute $83 billion to the U.S. economy within a decade, primarily through better academic performance (which, they predict, will create more productive citizens) and reduced traffic accidents caused by sleepy adolescents and parents. I wish I could time travel this data back to the 1980s and hand it to my shoulder-tapping father.

While debates continue about moving school start times and reconfiguring our culture's approach to work schedules, the takeaway from the research is pretty straightforward: get a sense of your own sleep rhythms (your chronotype), if possible adjust your schedule to fit that rhythm, and try to be consistent.

How do you figure out your chronotype? You probably already know what it is. Simply ask yourself, "Am I a morning person?" Of course, scientists have a more systematic approach. A 2018 study from the U.K. examined the chronotypes of almost a half a million people. It found that 27 percent of the population are definite morning types, 35.5 percent are moderate morning types, 28.5 percent are moderate evening types, and 9 percent are definite evening types. This was determined by using the well-known Morningness-Eveningness Questionnaire. The survey, which anyone can take (it's easy to find online), was developed in 1976 to get a sense of where people sit on the chronotype continuum. Its questions seek to discover when you are at your "feeling best" peak. (The test categorized me as a "moderately morning type," which seems exactly right.)

I know, I know. "Live to your chronotype" is the ultimate "easier said than done" advice. Many people do shift work, and many others have to be at work at 7 or 8 AM. Others need to get up early because they have young kids who require the presence of a semiconscious adult. But while we can't all craft our wake-up time to exactly fit our chronotype, the advice still has relevance.

Do not feel compelled to conform to the "early is always better" stereotype. You are not the Rock (unless you are, in which case, I loved *Furious 7*!). Everyone is different. Do your best, within the constraints imposed by the realities of your life, to find a rhythm that works for you.

Second, and perhaps most important, be consistent. Find a groove you can maintain.

But don't let the drive for consistency get too regimented. Relax, dammit! "Yes, it is better to be consistent with sleep and wake times," Dr. Panda told me. "At the same time it is also important to get enough sleep." If you are tired, sleep in a bit. Give yourself about an hour or so wiggle room. In fact, there is evidence that, if done only occasionally and only for a short time, a good weekend sleep-in can help a tired body recover. Consistency is key, for maintaining both health and a restorative sleep schedule. You shouldn't use the weekend sleep-in as a long-term sleep strategy.

But, as Dr. Samuels emphasized, it is all about finding your particular rhythm. When I asked him about all the sleep advice floating around, he said, "Don't be a fool. Don't apply someone else's routine to yourself. Don't adopt a pattern that is Tom Brady's. That may be completely contrary to what will work for you. There is no magic routine!"

So for now, relax about that wake time. You don't need to get up with Tim Cook at 3:45 AM.

6:31 AM—Check phone

Don't.

A 2015 study found that for most people in the U.S., their smartphone was the most important thing on their minds when they woke—not coffee, dressing, or even their significant other. It is no surprise, then, that 61 percent of us check our smartphones within five minutes of waking. About half of us check the moment we wake. For millennials, that number is 66 percent. There are two problems with this behavior.

First, you may end up texting or emailing or tweeting or Facebooking something you will deeply regret. When you first wake up, you are consumed by what is known as sleep inertia, a state of decreased cognitive function. Your brain is just getting warmed up and isn't yet working at full capacity. And it is the higher brain functions, the ones that will stop you from writing a cringeworthy text, that are the last to come online.

Second, checking your phone immediately may not be a great way to start your day. Compulsive phone checking is associated with anxiety and stress. Give yourself a bit of time before you dive into the sea of emails, direct messages, Facebook posts, and text messages.

In fact, make the decision the night before to place your phone in another room before you go to bed. This is probably the best way to avoid starting your day with an email kerfuffle or creating your own "covfefe" Twitter catastrophe.

6:35 AM—Brush teeth

The Egyptians first used toothpaste, likely applying it with a twig or finger, around 5000 BC. Toothbrushes have been around since roughly 3000 BC. The first bristled brushes—with bristles from the necks of pigs—emerged in China around AD 1600. Brushes looking somewhat similar to the ones we use today have been mass-produced since 1780. And while most North Americans didn't start brushing regularly until after WWII, spurred on by soldiers who brought the habit back from Europe, oral health is now a multibillion-dollar industry involving high-tech toothbrushes, a dizzying array of toothpastes, flosses, mouthwashes, and whitening products, and a host of fancy cleaning contraptions. But despite this long history and strong social commitment to our teeth, the evidence surrounding the benefits of many of our oral health practices is remarkably equivocal. Indeed, there is surprisingly little good research to support much of what we do to our mouth and teeth. Morning breath can no doubt have a serious impact on our personal interactions. But what should we really worry about when it comes to our morning routine?

I will tackle some of the science surrounding oral health at the end of the day, because that is when most experts agree you should do the bulk of your teeth maintenance. But the morning bathroom stop is the perfect time to consider the growing concern about fluoride and the fluoridation of our water.

This is a good example of the influence of fear and even misinformation. Some jurisdictions in the developed world, such as Calgary, Alberta, have decided to stop fluoridating their water supply. A 2015 study in the *Canadian Journal of Public Health* notes that "opposition to water fluoridation is witnessing a vigorous comeback." Celebrities such as Dr. Oz have given the anti-fluoridation advocates a voice. And it is a common theme on fake news websites. NaturalNews, one of the most notorious purveyors of health nonsense, featured an absurd article with the false headline "Hundreds of Brave Dentists Speak Out Against Water Fluoridation." And there has been a growth in anti-fluoridation advocacy groups that have pushed the circulation of theories—not supported by convincing evidence—that fluoride in water has lowered the IQs of children and caused various kinds of cancer.

As is so often the case, much of the public's angst associated with the fluoridation of our water supply seems to have started with bizarre conspiracy theories. Do a quick web search for "fluoridation" and you will find people suggesting that it is associated with communists, Nazis, and/or the Illuminati. One common theory is that governments put fluoride in the water to tranquilize the population. It is, the theory goes, a way for governments to produce a more subservient citizenry. A far-fetched idea, you might say, but a 2013 study found that 9 percent of Americans believe this to be true and another 17 percent are not sure. Those are pretty shocking numbers. Over a quarter of the population is open to the idea that their government has been systematically sedating the entire population for decades.

Another anti-fluoride meme declares that fluoridation started with Hitler and Stalin as a way to sedate the inmates of concentration camps. This stuff is patently absurd, both historically and scientifically, and is likely believed by only a small number of those concerned about fluoride in our water. Still, studies have shown that the mere exposure to conspiracy theories, even utterly bizarre ones, can distort public perceptions.

Conspiracy theories can serve as the original source of a health concern—in effect, giving life to a false claim that fluoridation is harmful. After the concern takes root, it can become associated with more mainstream and intuitively appealing ideas. In the case of fluoridation, these may include the belief that industry has had an inappropriate influence on the research or that, regardless of how efficacious fluoridation of the water is, it is an infringement of individual rights. Those are both ideas worth considering, but they are grounded more in ideology and personal branding than in evidence-informed decision-making.

And because the conspiracy theories about fluoridation are so bizarre and extreme—Hitler, mind control drugs, and a massive government cover-up are all involved— they are memorable. Even if you don't believe the core ideas, their existence helps to keep the general concern about fluoride in circulation and easy to recall. And, as we've noted, a significant body of research shows that the easier something is to recall and the more often you hear it, the more plausible it feels. Perhaps only a few believe the Hitler or mind control myths, but all the "noise" about fluoridation may nudge people to choose, for example, a fluoride-free toothpaste. And it appears that all the noise influences perceptions of benefit and safety. A 2015 survey of Americans done by the Centers for Disease Control and Prevention (CDC) found that a relatively large minority, 27 percent, believed that community water fluoridation had no health benefit, and only 55 percent believed it was safe.

But *should* we worry about fluoridation? Is there a vast conspiracy to use fluoride to anesthetize the population? Though research continues and we should regularly revisit the science, when it comes to an assessment of benefit and safety, the research is pretty darn consistent. As noted in the government of Canada's 2016 position statement on fluoride, "Community water fluoridation remains a safe, cost effective, and equitable public health practice and an important tool in protecting and maintaining the health and well-being of Canadians."

A 2018 study funded by the U.S.'s National Institutes of Health, involving 7,000 children aged two to eight and more than 12,000 older children and adults, also confirmed the substantial health benefit of community water fluoridation, particularly for children. While too much fluoride can have adverse effects, such as causing white specks on teeth, the amount added to drinking water for public health reasons is well within safe limits. But the adverse impact of stopping a fluoridation policy is greater. Indeed, in jurisdictions where fluoride has been removed, such as Calgary, the rate of tooth decay has increased significantly. Given this kind of data, it is no surprise that the CDC has declared community water fluoridation to be one of the top 10 most successful public health interventions in history.

When it comes to fluoridated water, the bottom line is clear: the evidence of benefit is solid and the evidence for harm is weak. If your community embraces fluoridation, be thankful and relax.

But let's bring it back to your morning brush. Should you use fluoride toothpaste? Here again the answer is clear: yes. A comprehensive review of more than 70 clinical trials involving over 70,000 kids confirmed the "benefits of using fluoride toothpaste in preventing cavities in children and adolescents."

In fact, from a health perspective, the presence of fluoride is really the *only* reason to use toothpaste. This point was made to me by Dr. Grant Ritchie, a dentist, writer, and vocal advocate for science-based approaches to oral health. He is, no surprise, a fan of brushing. "It is the mechanical action of the bristles removing the plaque that confers most of the benefit," Dr. Ritchie told me. "The main benefit of toothpaste is as a fluoride delivery system."

Despite this, many "natural" and "organic" fluoride-free toothpastes are on the market. Beyond possibly helping your breath smell organic-y, these products are likely totally useless. Indeed, I could not find a single study to support their use. One study, on charcoal toothpaste, noted a complete lack of good research. Dr. Robert Weyant, a professor of dental public health at the University of Pittsburgh, agrees with my assessment. "Don't fall for the naturalistic fallacy," he said. "These 'natural' products—and, by the way, 'natural' is never defined—are an example of marketing trumping science. There is no evidence they are effective. I would completely discount their value."

Dr. Weyant believes the public is subjected to a huge amount of misinformation about dental care. Taking care of your teeth is important, but, as is so often the case, it really comes down to focusing on the basics. "Eat a healthy diet, don't smoke, and brush your teeth twice a day with a fluoride toothpaste," he told me. "After that, everything else is secondary. If you live in a town with fluoride in the water, that is an added bonus."

(If you are wondering about what type of toothbrush to use, soft bristles appear to be somewhat safer—that is, they are less likely to cause injury—and there is some evidence electronic toothbrushes might be better for avoiding gum disease. But there isn't a ton of good evidence. A 2014 study found that "bristle design has little impact on plaque removal capacity of a toothbrush.")

It is worth noting that there is, in fact, no magic to brushing in the morning. The evidence tells us that you get the most benefit from brushing twice a day—again, primarily because of the delivery of fluoride—but no added benefit by brushing more often. And most experts agree on the importance of brushing at night, right before bed. "This is primarily because it allows fluoride to hang in your mouth, doing its job all night long," Dr. Weyant explained. So you don't *have* to brush in the morning, though there is an obvious morning-breath mitigation benefit associated with an AM scrub. So if you work with me, please brush!

6:40 AM—Check phone, again

Don't. You probably will. But you shouldn't.

It is remarkable how quickly smartphones have altered the human experience. Apple's iPhone was released in 2007, and since that date, smartphones have resulted in a staggering change in how we engage with the world. Some estimates suggest that the average person checks their phone over 100 times a day, looking at the device every 10 minutes or so. And we "touch" our phone—that is, perform some kind of function on it—in excess of 2,500 times a day. Crazy! A 2018 study found that people check their phones 80 times a day *while on vacation.* On average, we spend over four hours a day staring at these devices. That's 120 hours a month, equivalent to a pretty serious part-time job. About half of us check our phones—usually email first—before we even get out of bed.

So, for sure, most of you will check your phone at least twice (or maybe 10 times) between waking and breakfast. My guess is you are doing it on the toilet. If you believe 2016 market research, 75 percent of people check their phone while sitting on the toilet and about 40 percent read and send emails on the toilet. (Think about that the next time your coworker sends you an early-AM message.) This might explain why 19 percent of us have dropped our devices in the loo. It is also one reason, according to a 2009 study, that 95 percent of smartphones are contaminated with various forms of bacteria. (Think about *that* the next time someone asks you to take a picture for them!)

Despite the ubiquity of the early-AM phone check, there are good reasons why you shouldn't check your phone so frequently. I will tackle these later in our hypothetical day. For now, put down the damn phone.

6:45 AM—Step on the bathroom scale

People are stressed about their weight. A lot. One 2014 poll found that 21 percent of women worry about their weight all the time and 34 percent worry about it some of the time. Another 2014 survey of 2,000 Americans found that three in four adults "feel like they could always lose some weight." And this isn't just a North American phenomenon. A study from France found that about 45 percent of all Europeans are concerned about their waistline.

All this stress about our mass has led to a serious amount of (largely futile) dieting—and an enormous $220 billion weight loss industry. One estimate suggests that the average woman will go on 61 diets before age 45. And the number for men is increasing. The 2014 survey noted above found that 63 percent of American guys "always feel like [they] could lose weight." It is not surprising, then, that the weight loss industry is increasingly targeting men (and doing their best to stoke weight-loss anxiety). Companies like Weight Watchers have turned to male spokespeople, including retired football stars, to push their products on the segment of the population that has been, until recently, largely ignored by the dieting industry.

The social forces leading to our obsession with weight are many. They range from unrealistic media portrayals ("You need sexy abs for beach season!") to the marketing strategies of the weight loss industry ("This product will give you sexy abs by beach season!") to comparisons facilitated by social media ("Look at these sexy abs!") to a genuine desire to be healthier (which doesn't need to include sexy abs). Regardless of where the impulse originates, it is clear that we experience a lot of weight-related anxiety these days.

So, should you step on the scale? Does this machine help the weight loss and weight maintenance battle or will it just stress you out? Is this a good way to start your day?

These simple questions have resulted in much academic and public debate, as highlighted by the reaction to a decision in 2017 by Carleton University in Ottawa to remove the scale from the campus gym. "We don't believe being fixated on weight has any positive effect on your health and well-being," the manager for the university's health and wellness program said in defense of the controversial move. He argued that this scaleless strategy was "in keeping with current fitness and social trends" because weight "does not provide a good overall indication of health." The move sparked an almost immediate—and international—response. The *Daily Mail* in the U.K., to cite just one example, ran a headline stating, inaccurately, that the university removed the scales "because they are 'triggering eating disorders.'"

I've experienced a bit of scale backlash myself. When I tweeted about a study that explored how often people should weigh themselves, I got comments like "How about never!" and "A scale doesn't measure health!" and "Your self-worth isn't a number!" These widespread sentiments flow from the idea that stepping on a scale will hurt self-esteem, increase negative body image issues, and cause people to focus on weight instead of health. And these arguments against the humble scale have a great deal of intuitive appeal. A 2017 article in the magazine *Cosmopolitan* is typical. The title says it all: "I Threw Out My Scale and I've Never Been Happier." Countless similar articles can be found in pop culture. Do a web search on the phrase "throw out your scale" and you will get more than 300,000 hits—most of them blog posts and media articles embracing a "you are more than a number" approach to weight loss and body positivity. Indeed, this message is everywhere in popular culture. Actor Kate Winslet famously told Jimmy Fallon on *The Tonight Show* that she hasn't weighed herself in 12 years. "Top tip, it's a great move," the Academy Award winner said.

Many health professionals have also embraced this anti-scale ethos. For example, a 2018 CBC story on dieting included advice from a registered dietitian who suggests that when it comes to the scale, we should all, once again, "throw it out." (Bathroom scales must surely be clogging landfills throughout North America!) Why? Because, the dietitian asserts, it doesn't help and "the scale has a lot of power and hold on people on how they see themselves. They allow that number to determine their self-value, their self-worth."

But what does the science say?

In fact, there is plenty of evidence that, for most people, weighing yourself regularly can be a helpful weight loss and weight maintenance strategy. A 2015 systematic review of the evidence found that "regular self-weighing has been associated with weight loss." The authors of another systematic review went further, suggesting "self-weighing is likely to improve weight outcomes, particularly when performed daily or weekly." Another study followed 3,000 individuals for two years and concluded that "higher weighing frequency was associated with greater weight loss or less weight gain." A 2015 six-month trial that assigned half the participants to weigh themselves daily came to a similar conclusion: the act enabled "greater adoption of weight control behaviors and produced greater weight loss." Several investigations involving young adults and college students all conclude the same thing: the scale helps.

To be fair, these kinds of studies have their limits. Many simply found an association (perhaps people who like to weigh themselves are better at losing weight?), and some of the more controlled studies have found less impressive results. Still, taken as a whole, the data is pretty darn consistent and convincing. If weight loss or weight maintenance is your goal (and the latter should be a goal for most of us), stepping on a scale fairly regularly might be a good idea. As noted in the conclusion of a 2007 study by Butryn and colleagues at Drexel University: "Consistent self-weighing may help individuals maintain their successful weight loss by allowing them to catch weight gains before they escalate and make behavior changes to prevent additional weight gain."

Okay, so frequent self-weighing might be a helpful weight-management tool, but at what cost? What about its psychological and social implications? If stepping on a scale is going to make us all miserable, is it worth it? As emphasized by the Carleton University scale scandal, this seems to be the core issue. And I get it. With so many forces in popular culture that emphasize unrealistic weight loss goals, and with negative body-image issues on the rise, there are certainly reasons to be cautious about recommending a strategy that invites us all to reflect on our weight. Every. Single. Day.

The idea that stepping on a scale makes us unhappy or, worse, causes more serious psychological or health problems is a testable question. And, no surprise, researchers have tested it. The results? Despite all the pop culture finger-wagging, very little evidence supports the idea that self-weighing causes long-term psychological stress or body image issues, at least for adults.

The call to throw out your scale provides a great image that would seem to fit perfectly with the theme of this book—stop stressing out! But the evidence tells us that that image does not really reflect the science (which is an equally important part of this book!). You may be throwing out a useful tool that is not—at least according to most of the research—associated with significant psychological harm. A 2016 analysis, for instance, looked at all the data and came to the conclusion that, for the most part, stepping on the scale does not result in psychological harm. A 2014 clinical trial came to the same conclusion: "Self-weighing is not associated with adverse psychological outcomes," the researchers said, and is "an effective and safe weight-control strategy among overweight adults."

That said, it seems prudent to take much greater care with adolescents, particularly teenagers. There is evidence of an association between frequent self-weighing and issues of body dissatisfaction, likely because this is a subpopulation that, as the authors of a 2011 study suggest, "is often striving to meet thinness ideals and current beauty standards." But even among these young adults, self-weighing, when used carefully (such as considering vulnerability to body image issues), can be a useful tool. A 2015 study that examined the weight loss habits of almost 600 young adults concluded that "frequent weighing was associated with healthy weight management strategies, but not with unhealthy practices or depressive symptoms."

Given the evidence of benefit and the lack of evidence of significant harm for adults, why has this simple household device elicited such a strong reaction?

The idea of weighing ourselves is a pretty new phenomenon. For almost all of human history no one on the planet had any idea how much they weighed. Or, for that matter, what they *should* weigh. For most humans, anything north of starving was the hoped-for condition.

The earliest known weighing apparatus—an 8 cm balance scale made from limestone—was found in Egypt and dates from about 5000 BC. Small stone balance weights from around 2400 BC have been found in the Indus Valley in Pakistan. These simple devices were likely used by traders for weighing gold, grain, and other commodities. I think it fair to assume that humans were not turning to these early technologies to monitor their weight.

One of the first systematic attempts to collect data about human weight was undertaken by the Italian physician and professor Santorio Santorio (1561–1636), a colleague of Galileo who is considered the father of physiology. In an effort to gain a greater understanding of the human metabolism, he meticulously measured everything associated with, well, his metabolism. This included his food, liquids, urine, feces, and overall body weight. Basically, Santorio measured everything going in and coming out of his body. To assist him in his measuring goals, he used a "weighing chair"—an elaborate device that he sat in and that deployed a balance beam and a counterbalance—to monitor his weight before, during, and after various activities, including eating, excreting, and sex. (In case you are wondering, I could not find any data on how often Santorio had sex in this not very sexy machine.) He recorded this data for over 30 years. Truly the first quantified man.

The first scales that were accessible to the general public didn't arrive until the mid-1800s. They were clunky coin-operated machines that were often placed in public places such as train stations. Weighing oneself still wasn't an everyday activity, and these machines were presented more as a source of entertainment (Guess your weight!) than an essential health tool. Still, they quickly became popular, and hundreds of thousands of these pay-to-weigh contraptions spread across North America. In the 1930s, as a strategy to keep people stepping on the scales, the machines would spit out a slip of paper with your fortune or a picture of a movie star on it. (Given our current obsession with weight, the fortune-slip marketing ploy—which ties weight to how life will play out—seems like an ominous foreshadow.)

Early marketing for the first bathroom-style scales, which started to become more affordable in the 1940s, focused on health, not appearance. Around the same time, the idea of frequent weighing entered popular culture. "It's a national duty to keep fit. Check your weight daily," said a British advertisement for a war-era scale.

And since then, at least from the perspective of health care systems and public health authorities, the idea of weighing yourself frequently has never really gone away. That is, until the relatively recent pop culture call to throw the damn thing out. For example, the 1996 CDC National Diabetes Prevention Program's guidelines for weight maintenance recommends, "Weigh yourself regularly." (Interesting that the 2018 version is subtler, using charts and pictures to imply that you need to track your weight.) In the U.K., the National Health Service's recommendations on how to maintain a particular weight include "weigh[ing] yourself regularly so you can keep a close eye on any changes to your weight."

And as I write, my local public health department is in the middle of a cancer prevention campaign that encourages people to "maintain a healthy weight." The image that popped up on my Twitter feed in support of this initiative was a scale accompanied by the rhetorical question, "Did you know we can prevent about 673 cases of cancer in Alberta each year by maintaining a healthy weight? Learn how." The implication was clear: use a scale.

As Roberta Bivins, a historian of medicine at the University of Warwick, told me, the perspective of most public health officials was and still is: "Why would you get rid of a useful tool?"

Professor Bivins is one of the few scholars to dig deep into the evolution of our relationship with the scale. She says that good health has been a consistent theme in the marketing of the device, but by the 1960s there was a shift in the justification. "Selling scales as a thing associated with appearance and sex started to become more common. The health narrative is still there, but now it shares space with other concerns."

"The modern way is to weigh every day," declares a 1960 advertisement for a bathroom scale. "Every woman wants a trim figure, and there's no better or easier way to keep a check on your weight than with the regular use of a 'Mayfair' Personal Weigher."

The focus on appearance may explain why many people now view the scale as a wellness-destroying enemy. Over the past century, monitoring our weight has been pushed as a cultural virtue essential for our health and, increasingly, as something closely tied to our appearance. A lot of social significance, rightly or not, has been projected onto the number that is displayed by this small household appliance. And when appearance becomes the primary goal—rather than health—then disappointment, angst, and frustration become more likely endpoints, particularly given that research has consistently found that sustained weight loss is tremendously rare. (Another 1960s advertisement for bathroom scales embraced this reality, suggesting that their funky-looking products "give you the grim news with a grin.") Studies have found that people who are motivated to diet and exercise primarily by appearance—which is a dominant theme in the weight loss industry—are less likely to succeed and more likely to have negative attitudes about their body.

I believe self-weighing has, understandably, become tangled up in the tension between encouraging the maintenance of a healthy weight and concerns about weight stigma (that is, negative attitudes about individuals because of their weight) and distorted body image. The idea of often checking your weight feels like a path to an unhealthy relationship with your body, especially if appearance is your focal point. Intuitively, it feels like a mistake—an invitation to start the day with a bit of bad news. No wonder we've seen so many celebratory declarations to throw the damn thing out.

When it comes to talking about weight loss, these are strange and precarious times. On the one hand, the rise in obesity rates is an international crisis. More than one-third of the world's population is overweight or obese. There are, by far, more people who are overweight than underweight. And despite intensifying policy actions, the situation continues to get worse. A 2015 study published in *The Lancet* found that "no country to date has reversed its obesity epidemic," and a 2017 OECD report predicts that by 2030, fully 50 percent of the U.S. population will have obesity. Currently, obesity costs the world economy $2 trillion every year. It is a serious public health issue that demands an aggressive policy response. As a result, the topic of our weight is getting more and more attention. And a growing body of research explores how individuals and society should tackle this challenge—including using strategies like frequent self-weighing.

The causes of the obesity problem are ridiculously complex, involving dozens of interacting biological, behavioral, economic, and social factors. And while researchers are still teasing out the causal pathways, we do know that blaming and shaming those with obesity can be counterproductive. Weight bias—that is, discriminating against or stigmatizing individuals because of their weight—has emerged as a serious social concern that can, paradoxically, increase the risk of obesity. As a result, we need to take great care in how we talk about weight loss strategies, including the use of scales. Might pushing the idea of regular self-weighing contribute to weight bias?

I talked to Rebecca Puhl, deputy director of the Rudd Center for Food Policy and Obesity at the University of Connecticut, about the tension between dealing with the issue of obesity—which often involves encouraging people to watch their weight—and concerns about body image and weight bias. "The research is consistent," Puhl told me. "We know stigma is unhealthy. It is the enemy of public health. And it is linked to weight gain." Puhl should know. Having led many empirical studies on the topic, she is considered one of the world's leading experts on weight bias.

Puhl recognizes that for many of us the scale can be a helpful tool and that people who use it "tend to be more successful at maintaining their weight." Still, she worries that focusing on this approach may place too much responsibility on the individual. Might it encourage a "this is your fault" mentality and, as a result, heighten the problem of weight bias? "Knowing a number on a scale doesn't change the environment," Puhl notes. "It doesn't create healthy neighborhoods."

This is an important point. If we emphasize the value of monitoring our weight, it may create the impression that if the weight isn't coming off or staying off, we are, at some fundamental level, failing ourselves and our community. Given the complexity of the issue, that simply isn't true. And Puhl notes that this burden of responsibility can have a tangible effect. She points to studies that have found that shame about getting on a scale at the doctor's office can act as a barrier to obese women getting needed care. That is a terrible outcome.

At the same time, we shouldn't minimize the consequences of unhealthy weight gain. In this context, the scale's value seems clear, especially when we consider that a large percentage of the population doesn't even know how much they weigh. Indeed, people are not very good at guessing their own weight. A 2014 study found that weight misperceptions are extremely common, with 48.9 percent underestimating and only 6.8 percent overestimating their weight. Parents do an even worse job with their kids' weight. A 2017 study found that an astounding 96 percent of parents underestimated the weight of their overweight children.

On top of all this, as the average weight increases, so too does misperception about what makes for a healthy weight. A 2015 study of over 5,000 teenagers done at University College London found that almost half of the boys and a third of the girls who were overweight or obese "perceived themselves to be about the right weight." (The researchers also found a bit of good news: only a small portion of normal-weight teens, about 7 percent, thought they were too heavy.) The authors warned that this "lack of awareness of excess weight among overweight and obese adolescents could be a cause for concern."

This changing perception happens with adults, too. Just a few decades ago, most adults could correctly identify themselves as obese or overweight. In 1990, for example, about 56 percent of Americans were obese or overweight and 48 percent identified themselves that way. Today, more than 70 percent are obese or overweight but only 36 percent see themselves as being too heavy. (Given the number of people dieting, these numbers support the idea that they are doing it for appearance rather than because they view themselves as being overweight or obese.)

This misperception is likely due, at least in part, to a normalization of heavier weights. What people believe to be the ideal weight has increased significantly, suggesting a kind of social accommodation that has, as Professor Bivins told me, "shifted the 'normal' weight cues." Put another way, there are now fewer clues that we are gaining weight because almost everyone around us is gaining weight too.

As a practical example of how this is playing out culturally, Professor Bivins pointed to the evolution of clothing sizes in North America. Whereas there was once a generally agreed upon industry standard for women's clothes, since the early 1980s manufacturers have been left on their own to delineate sizes. Naturally, market forces began to influence their decisions and the era of "vanity sizing" was born. Manufacturers have come to use sizing as a way to attract customers through a strategy of sizing sycophancy. Today's size 8 dress is the equivalent of a size 16 in 1958. Research has consistently found that this marketing tactic works well, largely because—surprise, surprise—it makes us feel good. A 2012 study on why vanity sizing is so effective concluded: "Fitting into a pair of jeans labeled smaller than its true size can increase positive self-related mental imagery for consumers." But retailers can't be too overt. If the deception is obvious, customers might react negatively to the product, defeating the economic goal of the sizing strategy. Thus, market forces have allowed sizing to stealthily creep up in exactly a manner that, as Professor Bivins suggested, removes cultural cues about weight gain. No jean manufacturer wants to be the brand that you will forever associate with the day you said to yourself, "Gosh, I'm putting on weight."

In a world of confused messaging and marketing sleight of hand, stepping on the scale provides an objective measure of our weight fluctuations. The key, I think, is to not let the number popping up between your toes mean anything more than that. It is just a measure of fluctuation—a tool to be used to help you maintain a weight. Period. Don't let it be a source of comparison with others or some abstract social norm. It is not, as the now deeply clichéd pop culture messaging so often declares (rightly, it turns out), a measure of your attractiveness, your self-worth, or, necessarily, your health. Indeed, while having obesity or being overweight is associated with a range of health risks, a person with a larger body is not necessarily unhealthy. Your BMI is not an accurate measure of your health.

I know that disregarding the weight loss noise that permeates our society isn't easy. Whenever I return from a long scaleless work trip, I dread the first weighing. And I'm annoyed if I've put on pounds, even though I know this response is irrational and unhelpful. Paradoxically, for me, weighing myself frequently decreases the psychological impact. Indeed, a big reason to use a scale fairly frequently is that our weight fluctuates a huge amount, even in one day. Stepping on the scale once every few months will not give you an accurate picture of your weight or the direction you are trending.

To illustrate this point, I got naked and weighed myself every hour on the hour during the waking hours for an entire weekend. What did I discover? First, this is not a fun way to spend your weekend. Second, the daily ups and downs in my mass were significant. At 9 AM I weighed 179. At 11 AM, 181. At 3 PM, 182. And at 11 PM I weighed still more. The next day the pattern repeated. The variation was dramatic. It was almost as if I were unknowingly absorbing doughnuts and ice cream through my skin. I seriously had no idea where the weight was coming from.

If I weigh myself regularly—I'd say at least a couple of times a week but not more than once daily—the ritual flattens the negative psychological zap and gives me an appreciation of our natural weight fluctuations. It allows the scale to become a simple, useful monitoring device.

What, then, is the bottom line? I very much sympathize with the concerns about weight stigma and the potential for psychological harm. But our decisions about our weight loss and weight maintenance interventions should, as much as possible, be evidence-informed. We shouldn't let unproven assumptions—no matter how well intentioned—stifle the use of constructive tools. We should strive to not confuse the number on the scale with success, failure, or even our health status (though it is, of course, potentially relevant to our health).

So, should you step on the scale? If you think this will help you maintain a healthy weight (whatever that is for you) and it doesn't freak you out (and the research suggests it probably won't), then yes, step on the scale. Think of it as a tool, nothing more.

6:50 AM—Get dressed

Once upon a time I owned handmade custom leather pants. And I wore them. In public. More than that, I wore them onstage. They were tight. As I said, custom.

I was fronting a new wave band during that nanosecond of history when that wasn't a horrifyingly embarrassing thing to admit. It was agreed by everyone in the band that getting custom leather pants was a very new wavy thing to do. We each got a different color. Seriously. We did this.

Shortly after our first all-leather gig, a fan approached to tell me she'd enjoyed the set. I was feeling pretty rock star. But just as she turned to leave, she pointed at my crotch and said, "By the way, you can't wear underwear with those—especially tighty-whities."

This unsolicited advice led to an emergency band meeting. After carefully inspecting each other's rears in different lighting conditions, we all decided she was right. The underwear had to go. The underwear lines were simply not acceptable. Thus, all future gigs were performed commando. Sans undies.

Now, I've been calling these pants "leather," but in reality they were made from a leatherlike material. It was like wearing a tight-fitting plastic garbage bag. The stage was hot. I jumped around and sweated like a pig (which is actually a reference to smelting iron ore and not to pigs, who don't sweat except, perhaps, when they wear cheap plastic pants). We were five young men who rarely washed our clothes. You can do the math. When I unzipped my pants to pee, it was as if I were standing over a compost bin filled with old cabbage. I believe it was our light man who put an end to the leather madness. "Get rid of those effing pants or get rid of me!"

Why did I tell this story? Underpants are a fine idea, if comfort and the control of odor are a concern. So my recommendation is to wear them, as I have done every day since my leather pants fiasco. Surveys suggest that as many as 25 percent of us go commando some of the time and a whopping 7 percent live sans undies all the time. There are no health risks associated with this practice, but yuck. If this is how you want to swing through the day, fine. But please wash your pants if you are sitting next to me on the plane. To be honest, for some of us, the only big undergarment decision is: boxers or briefs?

There has been a bit of research—but not as much as one would expect given the intensity of the boxer vs. brief debate—that suggests snug underwear, such as the classic tighty-whities, results in higher scrotal temperatures and, as a result, a reduced sperm count. A 2018 study of men attending fertility clinics found that those who reported wearing boxers had a higher sperm concentration. Other studies have come to a less pessimistic conclusion, finding that the selection of underwear had no effect on time-to-pregnancy for couples aiming for that goal. If that isn't a goal, then wear whatever you find comfortable. If you are trying for a pregnancy, going with boxers might help a small amount, but the data isn't conclusive. It couldn't hurt.

To get the lowdown on the down-low advice for women, I went to my friend Dr. Jennifer Gunter, renowned gynecologist and author of *The Vagina Bible*. Her recommendation: ignore all the debate. "Women should wear the underwear they like and that is comfortable," Gunter told me. "Contrary to popular belief, the vulva and vagina are not so delicate that a scrap of fabric can lead to V mayhem." Although some science—but, again, not as much as one would expect given the attention the topic attracts—suggests cotton underwear can lower the risk of some infections (specifically, yeast vaginitis), Gunter is not convinced it is worth getting your undies in a twist worrying about it. "Underwear doesn't impact the vagina, which is internal," she said.

Of course, comfort, hygiene, and health are not the only considerations that go into our underwear decisions. Many of us are concerned about what our romantic partners would like us to wear, but most of us have it wrong. Heterosexual women prefer, by far, men in boxer-like undies. In fact, just 35 percent of women prefer tight-fitting briefs, which is the underwear of choice for 57 percent of men.

I could find no research on the attractiveness of skintight, leather-like plastic pants.

6:55 AM—Coffee

Drink up! Coffee doesn't dehydrate you. (Even though it is a diuretic, it still contains more fluids than you pee out, so in the end it is hydrating.) It doesn't lead to adrenal fatigue (mostly because adrenal fatigue isn't a real thing). Despite headlines to the contrary, it doesn't cause cancer. (Coffee is a carcinogen in the same way mineral oils are—probably not, unless you are a mouse and you've been force-fed massive quantities and, even then, probably not.) And coffee does not stunt kids' growth (a myth that seems to have been started in the late 1800s by C. W. Post, the cereal manufacturer, to sell his coffee alternative, Postum).

There is enough research to safely conclude that coffee—even a fairly large amount—is probably reasonably good for you and almost certainly not bad for you. In fact, coffee consumption has been associated with *lower* cancer risk and a host of other health benefits. (Keep in mind that this is often correlation, not causation, research, so we should take care not to overinterpret the conclusions. But recent research is starting to unravel the biological action of coffee and how and why it confers its health benefits. This kind of work, while not definitive, adds to the body of science and increases the confidence we can have in the "coffee is good for you" narrative.)

These findings probably don't come as a surprise to many of you. Still, many alternative health practitioners and some celebrity wellness gurus (I mean you, Gwyneth) continue to treat coffee as a harmful indulgence that needs to be avoided. Ignore. The science is not on their side.

(I hate tea, so no comment on tea. But in all likelihood it is fine too.)

7:00 AM—Breakfast

Everyone has heard that breakfast is the most important meal of the day. We should start our day with a healthy, filling meal. This sentiment has been repeated so often that I think it is fair to call it a nutritional truism.

I love breakfast so much that I've cultivated a host of annoying rituals to ensure my morning meal goes as smoothly as possible. The night before, I set out all the ingredients needed for my muesli, making sure that I have the ideal ratio of yogurt, berries, and nuts. I become irrationally upset if one of my kids disrupts this pattern. Heaven help the child that eats my blueberries.

But I wasn't always a breakfast believer. In university I would save my hunger for one massive late-afternoon lunch, buying the cheapest and most calorie-dense meal I could find—which was usually a chicken burger with fries and gravy. Obviously, university-age me could improve the nutritional content of the meal. But from a timing perspective, who has the right approach, the slacker university student or the uptight health fanatic?

Our veneration of breakfast is a relatively recent phenomenon. For most of human history, when we woke up we just ate whatever was available—and only if we were hungry (which was probably often). But in the late 19th century, consuming a healthy breakfast, pushed by people like John Harvey Kellogg, took on a near-religious vibe. It was linked with being hardworking, efficient, and morally upright. For individuals like Kellogg, good people didn't masturbate (he viewed that as one of the most evil behaviors) and they ate breakfast.

The combination of advertising, moralizing, and cultural momentum helped to elevate the status of breakfast. Many in the nutrition community hopped on the breakfast bandwagon, proclaiming it one of the most important parts of a healthy lifestyle. A WWII public health poster shows Daffy Duck being shaken senseless by a rivet gun, with a headline declaring, "You Can't Breakfast Like a Bird and Work Like a Horse!"

But this is one of those areas where conventional wisdom has it wrong—or at least somewhat wrong. Despite more than a century of breakfast proselytizing, the evidence surrounding the value of breakfast is surprisingly mixed. There certainly isn't any evidence to justify the "most important meal" crown. When it comes to weight loss, for instance, the data is pretty underwhelming. A 2014 study published in the *American Journal of Clinical Nutrition* randomly assigned more than 300 people to either eat breakfast or not eat breakfast. The result? No difference. The authors conclude that "contrary to widely espoused views, this had no discernable effect on weight loss in free-living adults who were attempting to lose weight." Similarly, a 2016 study from Canada of over 12,000 people also found that "breakfast consumption was not consistently associated with differences in BMI or overweight/obesity prevalence." On the other hand, a 2019 study from the U.S. that explored the effect of a public school breakfast program found that "the initiative had an unintended consequence of increasing incident and prevalent obesity."

There is, however, a body of evidence that has found an association between eating breakfast and a range of other benefits. A comprehensive review of the science done by the American Heart Association in 2017 concluded that although "breakfast consumption does not improve weight loss," there is some evidence that it "can contribute to a healthier eating pattern that leads to slight improvements in cardiometabolic risk profile"—that is, the chance of type 2 diabetes, heart disease, or stroke. Research has also found a connection between eating breakfast and concentration and school performance. A 2013 review of the literature concluded that there was "some evidence" that a quality breakfast was related to better academic performance. But, as is often the case with this kind of research, the authors also noted that the association "can be attributed, in part, to confounders such as SES [socioeconomic status] and to methodological weaknesses such as the subjective nature of the observations."

So, is there *any* truth to the "most important meal" mantra? I put the question to Dr. James Betts, a nutrition and metabolism researcher at the University of Bath. "I don't see great value in this question" was his blunt response. "Even if we assume breakfast is the least important meal of the day, we are still left wondering whether it is sufficiently important to consume it."

Dr. Betts has been involved in numerous research projects and clinical trials that explore the timing of nutrient intake and human health, and his work has led him to be very cautious of reifying any particular meal, including breakfast. He takes great care in describing what the evidence says, so I shouldn't have been surprised by my inability to coax a simple yes or no response. That said, it is clear he also doesn't think the morning meal is magical. Dr. Betts told me, "If we are considering weight loss and general health, then there isn't currently a consistent or convincing evidence base to suggest that having breakfast rather than skipping it will cause a positive or negative response."

Professor David Allison, a renowned nutrition and obesity researcher at Indiana University, agrees. "Whether breakfast is the most important meal of the day depends, first, on how one defines 'most important.' Important to whom? Important for what? It may be that breakfast is important to help some people feel comfortable later in the day or to enjoy their mornings. Beyond that, the benefits of breakfast per se are not crystal clear."

Dr. Allison noted that some studies suggest eating more calories early in the day can promote better metabolic health, but the data is not conclusive. "Other people seem to believe that breakfast is an essential meal of the day to promote cognitive alertness or good weight control," he told me. "It is not clear that either of those two things are true. Randomized controlled trials addressing both questions have not consistently demonstrated benefits."

So, the breakfast literature is far from definitive. One reason this hasn't permeated popular culture is that many of the studies do a poor job of representing the difference between causation and correlation. A 2013 study by the Nutrition Obesity Research Center at the University of Alabama at Birmingham found that much of the literature "has gratuitously established the association, but not the causal relation, between skipping breakfast and obesity." In other words, the published research makes it sound as though skipping breakfast causes weight gain, when, in fact, it is simply a correlation. Perhaps people who eat breakfast are already slimmer or lead healthier lives than breakfast-skippers.

Some people in the health research community view these kinds of association studies as nearly useless. "They are actually worse than not doing the study at all," Dr. Vinay Prasad told me. He was commenting on yet another breakfast association study, published in 2019, that concluded that eating breakfast was linked to better cardiovascular health. "They are incapable of getting closer to the truth than the preconceived belief we had at the outset," he told me. Dr. Prasad, who is an oncology and health policy researcher at Oregon Health and Science University, believes this is the case in part because these studies often rely on self-reporting (which is frequently unreliable) and are predisposed to "massive selection bias for 'significant' results that validate preconceived notions." Most important, he believes that our nutritional habits are so "intrinsically tied to the types of people we are, our social circles, and socioeconomic status" that it is near impossible to determine causation. Is it breakfast or some other behavior or environment factor that is conferring a benefit or harm?

Like me, Dr. Prasad isn't necessarily for or against breakfast. ("I am a fan of breakfast," he told me. "If by breakfast you mean coffee.") But he feels that breakfast is "massively oversold" and is concerned that this overselling confuses the understanding. "Ranking the importance of meals in general is a fool's errand. It's like trying to decide what was the most important urination of the day." (That would be the morning pee, but I get his point.)

Dr. Betts agreed that both the media and the scientific literature too often use inappropriate causal language. And, he pointed out, "hype and error can be introduced . . . by the press offices of research institutions and the journalists." It is certainly easy to find headlines that emphasize this inaccurate causation spin. "Skipping Breakfast Makes You Fat," "Skipping Breakfast Leads to Obesity," and "Breakfast IS Key to Losing Weight" are examples of newspaper headlines about a small, unpublished association study that did not uncover anything definitive about "makes," "leads," or "is key to."

As with all things associated with meals and food, culture plays a big role. The influence of proponents like Kellogg has endured. "People have strong, zealous beliefs about food and breakfast, often with an almost moralistic tone," Dr. Allison told me. "The idea of breakfast may relate to a feeling of righteousness for leading a disciplined life involving early rising." What we need, he said, is "more rigorous randomized controlled trials assessing the effects of breakfast."

The notion that breakfast is important *feels* right. And because it is easy to fall into a classic causation vs. correlation trap, it is no surprise that you can easily find authoritative voices that support this intuition-based conclusion. It is a good example of how, when making any decision, you should be sensitive to the nature of the evidence behind the push. And remember, just because something has intuitive appeal doesn't make it right.

Still, I think the university-aged me was making a nutritional error. For most people, eating a healthy breakfast is probably a sound decision, especially if, as Dr. Betts has noted in his work, you have a physically demanding job. While the best available evidence is not always that methodologically robust, most of it points in a pro-breakfast direction. But if you don't enjoy breakfast or you find a different eating pattern works for you, fine. Once again, the bottom line is to relax and do what is best for you.

7:05 AM—Milk

Before reliable home refrigeration, most North American homes received their milk by daily milkman delivery. There wasn't much selection. The milkman would simply leave the requested number of bottles outside your door. Usually it was whole milk, often from a local producer. But through the 1960s and '70s it became cheaper to buy milk at the grocery store, and the era of the milkman began to fade. Around the same time, the variety of milk options expanded. Skim milk, which was initially a commercial flop, was pushed as an inexpensive and healthy alternative. A 1942 article in the *New York Times* declared that it "will help win the war" because, in powdered form, it could be easily shipped to the troops overseas. Decades later, as a way to move product, the dairy industry marketed skim as a slimming option, a strategy that was helped by the snowballing demonization of fat. In the 1980s, many government entities, including the U.S. Department of Agriculture, officially embraced skim milk as the healthier choice.

But what *is* the right choice? There are now so many milk, and milk-like, products. What kind of milk should you have with your breakfast? Whole? Skim? Almond? Chocolate? And how about raw milk?

Let's start with the most basic question. Is milk good for you?

"Pus is one of the worst, for sure," my wife, Joanne, said without a moment's hesitation. "Disgusting." And she is a family physician. She has a broad knowledge of disgusting.

"Yep, it's a really, really bad word," agreed my son Adam. "Top of the list."

My daughter Jane pinched her face in revulsion. "Why did you even say that word to me? It's horrible!"

"Actually, the word 'moist' is the worst word in the English language," my son Michael calmly interjected. "But the word 'pus' is obviously one of the grossest."

This family debate on the world's worst words was ignited by a blog post on milk. "White Poison: The Horrors of Milk" was the title of this impassioned piece on, well, the *horrors* of milk. The author starts by telling us that "the pus, blood, antibiotics, and carcinogens in milk—and the chronic fatigue, anemia, asthma and autoimmune disorders caused by milk consumption—do nobody good." That is pretty hard-core. *Pus?*

These extreme anti-milk sentiments can be found everywhere. Anti-milk documentaries, like *Got the Facts on Milk?*, and anti-milk books, such as *Milk: The Deadly Poison*, suggest that milk consumption is associated with a range of ailments, including cancer and heart disease. Many celebrities have publicly declared their disdain for all things dairy, among them Tom Brady, Jessica Alba, and several members of the Kardashian army, claiming it is both bad for you and an enemy to the waistline. (While the Kardashians have shunned cow's milk, Kim is fine with drinking her sister's breast milk as a way to improve her skin.) There are also many myths, like the idea that drinking milk will make mucus worse when you have a cold. (That is a medieval myth with no scientific backing.)

It is no surprise, then, that people are drinking less milk. A 2017 government of Canada study found that "Canadians' dairy choices are trending away from traditional milk and processed or high-fat products towards lower fat dairy sources and dairy alternatives." In my home province of Alberta, milk consumption declined 21 percent between 1996 and 2015. In the U.S., the intake of milk dropped 33 percent between 1970 and 2012. There are many reasons for this shift, including demographic changes and the increasing popularity of vegan diets. But health trends, and the growing perception that there is something wrong with milk, are clearly playing a role.

As with many diet and health beliefs, there is a political dimension to the anti-milk trend. Many people believe that the dairy industry has had an inappropriate influence on past food policy. There is evidence that the industry manipulated not just past dietary guidelines—which often recommended consuming a large amount of milk—but also some of the relevant research. In addition, people may have animal welfare concerns. And so it is easy to understand why an anti-milk position can *feel* correct. It feels like a noble and ideologically appealing position. But we should take care not to conflate frustration with the role of industry, or concern about animal welfare, with what the evidence actually says.

So, does the evidence support these fears?

In Canada, the antibiotic concern is easy to answer. Dairy farmers use antibiotics to treat sick cows, and when the cows are receiving the drugs, they must be taken off the milk production line. Antibiotics are also used—and some would say that they are used too often—for prevention when cows are not producing (this is often called "dry cow therapy"). But regardless of how and why they are used, the antibiotics must be completely cleared from the cow's system before she can be put back in production. The farmer is required to test the milk to ensure no traces of antibiotics remain. If antibiotics are detected in any of the milk from the farm, all the contaminated milk is dumped and the farmer faces a fine. Further, the use of growth hormones in the milk production process is illegal in Canada.

In 2015 the U.S. FDA published data after sampling milk for drug residues. Out of the 1,912 samples, only 15 had detectable residue. Drug residue in milk is a rare phenomenon, and the FDA's study helps to confirm the safety of the milk supply.

There are challenging issues associated with the growing problem of antibiotic resistance. We certainly need to do more to encourage a reduction in the use of antibiotics in agriculture. But that is a different issue from the concern that there are antibiotics in the milk we drink.

The fearmongering about pus and blood is, no surprise, misleading—though, to be fair, it depends on your definition of pus. The milk industry is highly regulated, and independent inspectors often test the product. Yes, there can be a small number of animal cells in milk (just as there are human cells in human breast milk). These cells are usually leukocytes (white blood cells). When living organisms are involved, whether plant or animal, you are going to find cells. But there isn't any pus or blood in the manner portrayed by the anti-milk crowd. That is, there aren't significant amounts of red or white blood cells floating around in milk products in a manner that would make them unsanitary or unsafe. And let's be honest, this language is used more for its visual and rhetorical power—as highlighted by our family debate—than as a serious argument against milk. Unless you are a vegan, humans consume cells from animals at almost every meal.

How about milk's purported health benefits and harms? This is where it gets complicated. Many studies support the consumption of milk. For example, a 2016 Danish review of the research concluded that the evidence "supports the intake of milk and dairy products" and that these "may protect against the most prevalent chronic diseases." A large 2018 multinational study published in *The Lancet* followed more than 130,000 people for nine years and found that a moderate level of dairy consumption "was associated with lower risk of mortality and major cardiovascular disease events in a diverse multinational cohort." Other studies have come to a more underwhelming conclusion, such as a 2016 review of 21 studies that found "no evidence for a decreased or increased risk of all-cause mortality, coronary heart disease, and stroke associated with adult milk consumption." In other words, milk was found to be not especially bad or especially good for you. (The authors of this review note that publication bias—the tendency to publish more positive results, especially when industry funding is involved—created a degree of uncertainty.)

After wading through every study, academic commentary, and independent nutrition guideline I could get my hands on, some clear themes emerged. Milk is not a superfood. Over the years, its benefits have likely been oversold. Industry pressure and marketing are partly behind that. But it is also not a poison. And it is a good source of many nutrients.

Despite the confused messaging in both popular culture and the academic literature, what I heard from many nutrition experts I consulted was surprisingly consistent.

I asked Professor Rhonda Bell, an esteemed expert in human nutrition at the University of Alberta, to give me one sentence on milk. Just one. She ended up providing me with a thoughtful two-paragraph preamble that analyzed the literature and then, feeling compelled to respond to my request, concluded with this nicely succinct summary: "I guess my one-liner is, 'It's pretty good.'"

Dr. Jayne Thirsk, senior director of Dietitians of Canada, agrees. She told me cow's milk is a "good source of an array of key nutrients, including standardized amounts of vitamin D," and, in her estimation, dairy has been "unfairly maligned."

Dr. Daniel Flanders, a pediatrician who specializes in nutrition and obesity, agrees that dairy products "are nutrient-dense food choices which happen to be a good source of calcium, vitamin D, fats, and protein." But Dr. Flanders also agrees that "dairy products are neither magical nor essential." Still, he says, "people ought to feel comfortable choosing dairy products as part of their balanced diet."

The U.K.'s National Health Service provides a concise review of the situation: "Milk and dairy products, such as cheese and yoghurt, are great sources of protein and calcium and can form part of a healthy, balanced diet."

In sum, milk should not be viewed as some supernatural beverage required by all. If you can tolerate it (a significant portion of the population is lactose intolerant), there is no health reason to avoid milk.

That said, there are some kinds of milk that, despite building cultural momentum to the contrary, should be avoided. I'm talking about the growing craze of consuming unpasteurized, or raw, milk.

The food blogger Vani Hari, also known as the Food Babe, has stated that when it comes to milk, she believes "that 100% grass-fed raw dairy is the best choice." And Gwyneth Paltrow has suggested that everyone should go on a raw goat milk cleanse to rid our bodies of parasites—which according to Gwyneth's "research" team, we all have. Who knew?

These kinds of celebrity endorsements, which often include claims that raw milk is a superfood that will transform your health, have helped to drive a growing market for raw milk. There are now raw milk advocacy groups throughout the world calling for the deregulation of the milk industry. And these efforts have worked. The popularity of raw milk is increasing. In the U.S., a growing number of jurisdictions are allowing the sale of raw milk. The underlying argument, as put forward by the Food Babe, is that "raw dairy products are 'alive' and have all of their probiotics, vitamins and enzymes intact." In other words, raw dairy is healthier because humans haven't interfered with it.

(It is curious how our culture can simultaneously sustain such polarized milk-focused niches. One paints milk as a superfood and elixir of health, the other as a pus-filled poison to be avoided at all costs. The former camp embraces the "natural" qualities of milk, including the many biological impurities. The latter paints those same qualities as harmful.)

Despite the intuitive appeal of the raw milk position, it is spectacularly incorrect. A 2017 study by the Centers for Disease Control and Prevention came to the striking conclusion that while only 1.6 percent of Americans drink raw milk, it accounts for 96 percent of milk-based illnesses. Put another way, in the U.S. raw milk causes 840 times more illnesses and 45 times more hospitalizations than pasteurized milk. And there are more dairy-associated illnesses in jurisdictions that allow the sale of raw milk than in those that restrict its sale. The Food and Agriculture Organization of the United Nations, the U.K.'s National Health Service, Health Canada, and the CDC, to name but a few, have all warned against drinking raw milk. The CDC, for example, notes that drinking raw milk could make you very ill or kill you. Yep, death is a recognized risk associated with this nutrition trend. In 2017, two people died after getting sick from eating artisanal raw milk cheese from upstate New York.

This is precisely why many jurisdictions throughout the world have banned or tightly regulated the sale of raw milk and require the pasteurization of all commercially distributed dairy products. Louis Pasteur, the celebrated French microbiologist, conducted the first pasteurization tests in 1862. Commercial pasteurization was introduced in 1895 and radically improved the safety of milk, allowing for the storage and transportation of dairy products. Though pasteurization is often portrayed as a complex, highly industrialized process, it is really just the straightforward heating of milk to kill harmful bacteria. There is no evidence that the process significantly reduces the nutritional value of dairy products in any meaningful way. Indeed, some countries use an ultra-high temperature (UHT) pasteurization process that allows milk to be stored without refrigeration. In France, for example, more than 95 percent of the milk has undergone UHT pasteurization.

As noted, advocates believe raw milk has important health benefits. A common claim is that raw milk reduces allergies and asthma. The theory is that exposure to the "healthy" bacteria in raw milk will give a boost to the immune system. Raw milk supporters point to observational studies that suggest that kids who drink raw milk have lower levels of asthma and allergies. However, these studies are far from conclusive. It is unclear whether it is the bacteria in the milk or some other product, substance, or action that confers a benefit. That is, raw milk consumption might simply be correlated with some other behavior that reduces allergy and asthma risk. (That correlation/causation dilemma again!) Regardless, the evidence is weak. As noted by a 2015 scientific review done by the New Zealand government, "The evidence does not warrant a recommendation promoting raw milk consumption in children who are at risk of allergic disorders"—especially since "exposure to raw milk in this age group is particularly risky given the susceptibility of this age group to infection." This was also the conclusion of a 2015 University of Wisconsin–Madison review of the evidence: "Raw milk is not inherently safe and carries a significant food poisoning risk with its consumption. There is no evidence that raw milk has any inherent health or nutritional benefits."

But even if raw milk did have the health and nutritional attributes claimed by its advocates—and the body of evidence doesn't support those claims—the benefit of those differences to our health would be so marginal, if measurable at all, that it could not justify a shift away from the pasteurized product. Do slightly more vitamins, minerals, and enzymes really matter? The answer is no.

In reality, the science is largely irrelevant to the raw milk movement. Drinking and advocating for raw milk have become a political declaration. It is an aesthetic act—a fashion statement—as much as it is a nutritional choice. Raw milk has become emblematic of food libertarianism. Food that is natural and devoid of bureaucratic interference is seen as inherently better, regardless of what the evidence says about the actual risks and benefits. Drinking raw milk represents the embrace of freedom of choice over government-mandated restrictions. It is a small act of rebellion. And this framing has proved very effective. There is evidence that arguments grounded in freedom of choice ("food freedom") have more resonance with the public than those based on public health concerns. Especially if someone is new to this topic, the idea that people should have the right to choose may have more traction than arguments about potential harm to the public.

The raw milk debate demonstrates how a vocal, politically minded few can spread misinformation in a way that can skew the decisions of many consumers. The popularity of raw milk is on the rise, and this is likely due to the spread of unsubstantiated claims of health benefit and the conceptually incoherent natural-is-better narrative. And because these claims are often wrapped in an intuitively appealing argument focused on freedom of choice, the idea of raw milk has had a broader appeal than the science—or, even, common sense—would seem to justify. Linking dodgy science to an attractive ideological stance is a surefire way to slip dodgy science into the public discourse. (Here is an example of this approach that I come across almost every day: Big Pharma is evil, so therefore pseudoscientific alternative medicines, such as homeopathy, are effective. If you don't believe they are effective, you must therefore support Big Pharma!)

There are clearly harms associated with drinking raw milk, but we need to be careful not to overplay what those risks mean to individual consumers. As the CDC study highlights, at the population level, it is easy to see the harm associated with drinking raw milk compared with pasteurized milk. But at the level of individual consumption, the chance that a single raw milk product will make you seriously ill is relatively small. (The risks for children and those with compromised immune systems are likely higher.) If someone offers you raw milk cheese, by all means, try it. The risk of illness is obviously increased, but perhaps not to the point where your enjoyment of cheese should be missed.

So what's the answer? Consuming raw milk is a straightforward risk vs. benefit calculation. Raw milk has no clear health benefits and empirically demonstrated harms. Yes, you could spend a lifetime drinking raw milk without experiencing any adverse effect, as people from rural communities have done. But that doesn't make drinking raw milk a sensible decision. You could drive around without a seat belt for years, perhaps a lifetime, without an adverse event. But seat belt laws are good social policy—the CDC estimates that seat belts have saved 255,000 lives since 1975 in the U.S. alone—and their consistent use is a wise personal decision.

The science review done for the New Zealand government provides a nice summary of this discussion: "The claimed health benefits of raw milk compared with pasteurised milk are for the most part not backed by scientific evidence, making the risk/benefit ratio very high for this food product, particularly among the vulnerable groups."

Don't let your decision-making around raw milk be clouded by ideological spin. Fact: pasteurization saves lives.

Before we leave the morning milk decision, a quick word on skim and low-fat dairy. Despite decades of advocating for low-fat options, the scientific community has moved toward a more forgiving position when it comes to fat content. This has been led by emerging evidence that dairy fat probably isn't that bad for us. Indeed, there is at least some evidence that points in the opposite direction. A 2013 study published in the *Scandinavian Journal of Primary Health Care* found that the intake of dairy fat was associated with a *lower* risk of obesity and the intake of low-fat dairy was associated with a *higher* risk. A 2013 meta-analysis of the research determined that there was no evidence to support the suggestion that skim milk is healthier. And a 2016 review of the evidence concluded that "there is no apparent risk of potential harmful effects of dairy consumption, irrespective of the content of dairy fat, on a large array of cardiometabolic variables, including lipid-related risk factors, blood pressure, inflammation, insulin resistance, and vascular function."

We should be careful not to overinterpret this data—much of which is observational. But, as an emerging body of evidence, it is convincing. It suggests that when it comes to the fat content of our milk, we should simply—yep—relax and follow our taste buds. Whatever fat content you enjoy is probably just fine.

7:15 AM—Vitamins

Vitamins: A Haiku

Vitamins, worth it?
A huge industry says yes.
Evidence says no.

Insurance, you say!
Sorry, the pills do not work.
Balanced diet, please.

Science is ignored.
Still they push, for the money.
Look! Expensive pee.

I realize that a haiku is not a very satisfying way to analyze a complex topic, but I figured it was worth a shot. Although intriguing research and academic debate continue in some areas, such as with vitamin D, overall there is a lack of evidence to support consuming vitamins and supplements. As well, studies have shown that many supplements are often contaminated or do not contain the ingredients on the label. Yet this largely science-free industry has a multibillion dollar market.

The bottom line: in general, you can ignore the supplement boosters. Unless you have a clinically identified need—and by "clinically identified" I don't mean by a bogus test offered by alternative practitioners—you should strive to get your vitamins from the food you eat.

7:45 AM—Kids off to school

On my way to work, I cycle past our neighborhood elementary school. It is a classic North American suburban school: a one-storey, flat-roofed utility box with a small gym stuck onto a corner. All four of my kids went there. And decades earlier, my wife and her siblings went there too. Much has changed since my wife walked the halls, of course. The school now has a funkier playground than it did in the '70s. And there are computers and smart boards (whatever those are) and worries about Wi-Fi waves.

But perhaps the biggest change is the traffic at the start of the day. If I'm riding by at the moment of peak kid drop-off, I will need to navigate through a throng of SUVs pulling in, pulling out, and making ill-advised U-turns.

Understanding why more and more parents drive their kids to school is far from straightforward. One study from 2017 found that convenience was a dominant reason: parents said that driving saves time. But this is only because a significant portion of those parents said they wouldn't let their kids walk without them. In other words, it saves time only because the parents aren't spending time walking with their kids. That same study also found that, surprisingly, distance didn't really matter. Although a long distance could factor into the decision to drive, shorter distances did not. The researchers concluded: "Distance does not affect the decision to take the car, which supports some research suggesting that parents drive their children regardless of proximity."

So what is behind much of the decision-making around how to get kids to school? Fear, primarily of traffic and degenerate strangers. Let's dissect these issues, starting with the "stranger danger" concern.

A 2009 study found that, for parents who said safety was a primary issue, fear of strangers was by far the dominant concern. These parents "reported much greater concern about danger to their children from strangers than they did about traffic concerns, and 75 percent did not allow their children to walk to school without adult supervision." A 2010 study interviewed dozens of Canadian parents about their decision to drive their kids to school, and it too found anxiety about strangers to be a big theme. Parents talked about the fear that their child "might meet a stranger on the way to school" or "some freak person" or feared "some wacko grabbing her and putting her in the back of the truck." And a 2015 survey done by the Pew Research Center in Washington, D.C., found that their being kidnapped was the third-biggest worry parents had about their children, behind being bullied and concern about mental health issues.

In reality, the risk of a child getting abducted by a stranger is astoundingly small. A 2011 study from Canada found that out of 46,718 children reported missing, only 25 were abducted by a stranger. And that study defined relatives and close friends as people *outside* the immediate family—meaning that many abductors were not true strangers. Indeed, a 2003 study found that out of 90 incidents that the authorities listed as "stranger" abductions, just two of those children were abducted by persons who were not a relative or a close family friend. A 2015 position statement authored by 19 academic experts on outdoor play concluded that the chance of a child being abducted by a stranger is about one in 14 million. This makes the possibility so fantastically remote that, in terms of daily risks, it can be categorized as "simply not going to happen."

But those rare odds are difficult for many of us to comprehend. Indeed, a core message of this book is that we are all pretty bad at assessing risk. So let me frame the remoteness of the abduction risk another way. Let's say, for some strange reason, you *wanted* your child to be abducted. Author Warwick Cairns has calculated that in order to ensure this event would occur, you'd need to leave your child alone on the street for 200,000 years. And your kid would most likely be returned within 24 hours. If the goal were to have your child abducted *and* murdered (yes, I know, this is a fantastically ghoulish thought experiment), you'd need to leave your child alone outside for several million years.

Mariana Brussoni, one of the authors of the above-noted position statement, is a professor of public health and an expert on outdoor activity and child safety. Her research has highlighted the health and development benefits of unsupervised outdoor activities, such as climbing trees, swinging on swings, exploring nature, and, yep, walking to school. Her 2015 systematic review of all the available evidence concluded that we need to support and promote "risky outdoor play for healthy child development."

I caught up with Professor Brussoni while she was in London. She was in the middle of what could be described as a mini world tour, visiting various research institutions to learn about the latest science exploring the benefits associated with independent, unsupervised play. I asked her what she feels is the primary reason parents have become so hesitant to let their children walk to school.

"It is complex," Brussoni admitted, recognizing such factors as convenience and timing. But in her experience, "what underpins everything is anxiety about the 'white-van man.'" She quickly added that parents are not making a rational and straightforward risk assessment. "This is tied to the broader social issue of kids' independence," she said. But, she told me, in contrast to what the statistics tell us, when it comes to abductions by strangers, "parents have the perception that it happens a lot." She explained that a number of cultural forces create this misperception.

The stranger-danger fear is stoked by news media portrayals of abductions—which, a 2011 study demonstrated, disproportionately emphasize stories of young white girls being "snatched from their middle- to upper-class homes by male strangers." Those news stories frame abductions around themes associated with family values, evil outsiders, and community cohesion. In addition to injecting elements of class and race, this kind of storytelling creates vivid imagery that likely makes the danger easier to imagine, and as a result it feels more real and present. Popular TV shows like *Bones*, *Criminal Minds*, and the *CSI: Crime Scene Investigation* and *Law and Order* franchises add to the impression that the world is populated with cannibalistic serial killers.

A growing body of research suggests that exposure to this kind of material can distort the public's perception of crime and the criminal justice system, including increasing the public's fear. Research done at the University of Pennsylvania found that the portrayal of crime on TV was associated with the public's fear of crime, regardless of the actual crime rates—which are, in fact, much lower than when today's concerned parents were children.

In addition, social media discussions of child abductions and the widespread use of Amber Alerts, those public service announcements designed to assist the rapid recovery of abducted children, may also heighten the perception that stranger danger is a significant problem worthy of parents' vigilance. This emergency response system was inspired and named after Amber Hagerman, a girl who was abducted and murdered in Texas when she was just nine years old. But despite this tragic backstory—which gives each Amber Alert an added poignancy—several studies have questioned the utility of Amber Alerts. A 2007 study by Timothy Griffin and Monica Miller, for example, concluded that while Amber Alerts have resulted in some dramatic successes, in the aggregate they do little to help the return of children abducted by strangers. The authors call the system "crime control theater" and suggest it is a socially constructed solution to a socially constructed problem.

One could reasonably argue that Amber Alerts are worth it if they help to save even one child—a point often made by the system's advocates. As a parent, I certainly get that perspective. In a minority of Amber Alert situations, they may assist in locating children who are abducted by a family member—which is by far the most common scenario (though a

2016 study found that the alerts usually have no direct role). But we also need to recognize that the system comes at a cost. It helps to generate and legitimize unrealistic fears about abductions. And, as we will see, this fear may have a wide range of adverse social consequences.

It is worth pausing here to emphasize two points. First, I sympathize with the "white-van man" concern. Having a child abducted and harmed is a powerful, primal, and perhaps evolutionarily driven dread. Some scholars have suggested that the current fear of the amorphous stranger is a modern embodiment of the "bogeyman" of myth. In a world of uncertain dangers, the idea that creepy strangers are lurking nearby can stand in for a host of undefined anxieties. Second, this fear is not, as we've seen, rational. It is entirely understandable that parents are not making dispassionate assessments of the statistical probabilities of harm. We are not robots. There is a bit of evidence that many parents recognize that their fear of harm to their kids is far out of proportion to the actual risk, but, regardless of that recognition, their decisions concerning the independent mobility of their children are governed by fear, not facts.

As is the case with almost every parent, I have, on several occasions, experienced that overwhelming, blood-to-the-brain, breathless panic that accompanies the belief that your child is gone. Years ago, for example, we lost our youngest, Michael, in a tourist-packed Edinburgh Castle. Now, if you must go missing, you could pick a worse location. But Michael, who was five at the time, seemed to have simply vanished. Poof. This immediately ignited my worst fears. Abduction. Harm. Gone forever. After a frantic search that involved the recruitment of castle staff, we found Michael calmly sitting by an information booth. Though we still don't know exactly what happened (the malevolent meddling of a medieval ghost remains an active theory), he was fine and, if we were capable of rational reflection at the time, we could have predicted this would have been the likely conclusion of the incident. But fear isn't rational, especially when you are talking about your kids.

I once also left one of my daughters—then a newborn—in a locked stairwell. But that's another story.

When it comes to stranger danger, there seems to be a self-reinforcing cultural feedback loop at play. Professor Brussoni called it a "cohort effect." As fear (and the spread of misinformation) causes more parents to drive their children to school, it is increasingly viewed as a parental norm and, as a result, *not* driving your child is viewed as abnormal, as bad parenting.

This social "norm" is what made Lenore Skenazy a really, really bad parent. How bad? She has been dubbed "America's worst mom." It all started in 2008, with her decision to let her nine-year-old son travel home alone from a department store on the New York City subway. She wrote about the decision in an article published in a local newspaper and this led to a barrage of parent-shaming.

"It was so strange," Skenazy told me, still exasperated by the experience a decade later. "I wrote that article and two days later, I'm being asked to defend myself on the *Today Show*, Fox News, and NPR. It was everywhere." Even though her son got home perfectly safe and was in fact energized by his independent trek from Bloomingdale's, the dominant reaction was that Skenazy was clearly a negligent parent. "The message was: 'How dare you put a kid's safety in danger,'" Skenazy told me. "It is fuelled by hysteria. Fear, death, worry, dread. And I hate decisions based on hysteria."

Other parents have been subjected to more than just aggressive shaming. In 2016 a Winnipeg woman was interrogated by Child and Family Services for letting her children play, unsupervised, in her backyard. In South Carolina, a mother was arrested for letting her nine-year-old play alone in the park across the street from where she worked. In Florida, parents were charged with neglect after their 11-year-old son played in their yard for 90 minutes. In 2017 a Vancouver man was ordered by

British Columbia's Ministry of Children and Family Development to stop letting his kids take public transport to school.

Cases like these have helped to create a parenting norm that requires children to be under almost constant direct supervision. And if you aren't supervising your kids—that is, if you let them play independently or walk alone to school—you are engaging in morally unacceptable behavior. A 2016 study found that people judge parents more harshly if they leave their children alone intentionally (for example, because of a need to work or, worse, just relax) as opposed to accidentally. More interesting, the researchers found that "the less morally acceptable a parent's reason for leaving a child alone"—such as work, relaxation, or, gasp, an affair—"the more danger people think the child is in."

This highlights the great power cultural norms have over parents' decision-making. The choice to drive your kids is motivated not only by an irrational fear of abduction but also by the more immediate concern (perhaps unconscious) of peer judgment. "There is a lot of parent blaming in this area," Professor Brussoni told me. "Whatever happens to your kids is your responsibility and reflects on you, especially mothers. And social media is there to help pass judgment. As parents, we are constantly being judged."

As someone who has been very publicly judged and shamed, Lenore Skenazy couldn't agree more. And she feels things are just getting worse. She notes that technology has made it easier to monitor one's kids—there are companies selling a variety of wearable devices that allow parents to remain in constant contact with their offspring—and so the expectation that you *will* monitor your kids continues to intensify.

Naturally, the advertising surrounding this emerging market—which Skenazy has called the child-safety industrial complex—serves as another cultural force driving parents' fears. To cite just one example, the website for GPSprotectsourkids.com (the internet address says it all) declares that kid GPS trackers are a must, because "today the world presents children with many challenges" and "the threat to their safety is real from abduction, sexual assault, school shootings to bullying." The site also tells parents that 76 percent of "abducted children [are] killed within 180 minutes of abduction"—a completely misleading statement. And the implication, of course, is that as a parent you must be constantly vigilant. Evil lurks. And you must be able to react quickly!

I think it is fair to say that, despite this kind of marketing and persistent parental fear, the stranger-danger myth has, from a research perspective, been thoroughly debunked. But what about the traffic concern?

Unfortunately, children do get hurt by traffic while walking to school. Exact statistics are hard to come by, but in my home city of Edmonton—which is a metropolis of over one million—there were two school zone pedestrian injuries in 2016. A 2014 study done in Toronto, a city of approximately three million, found that over a 10-year period, 30 school zone traffic accidents involving children resulted in serious injuries. There was one fatality. So these tragic events do occur, but they remain, in the big picture, relatively rare events. Indeed, a compelling argument can be made that, from an injury prevention perspective, it is safer to have your child walk to school than to drive them. As suggested by University of Toronto researcher George Mammen, "Evidence shows that children are more likely to be harmed in a car accident compared to walking to school."

Driving is, without a doubt, one of the most dangerous things humans do. In the U.S., car crashes kill approximately 35,000 people every year, and for the ages 5 to 24, the ages when kids would be walking to school, they are by far the leading cause of injury-related death. To be fair, this is partly because there are more people generally and more of them driving more miles than ever before, which increases the statistics compared with, say, walking as a mode of transportation. But even when we compare deaths per miles traveled, driving does not come out as a significantly safer mode of travel.

The key point is that there is no evidence that driving your children to school is an inherently safer approach. (If safety is your dominant concern, perhaps you should considering flying your kid to school on a commercial airline, as flying is probably the safest mode of transportation. But then you'd have to deal with airport security every single morning. Not fun.)

As with the stranger-danger concern, a powerful feedback loop is at play here. With more kids being driven to school, parents' concerns about traffic intensify and they may believe that walking has become less safe. And so they drive their kids—which compounds the traffic problem. And round and round it goes.

It is no surprise, then, that a growing body of research tells us that traffic near schools is a problem. Congestion is unsafe, and it's made worse by bad driving. A 2016 study from the University of Toronto analyzed the traffic around 118 schools and found dangerous drop-off behaviors at 88 percent of them. What's more, an estimated one-third of all collisions occur within 300 meters of a school. Despite the parental-concern-created car congestion, then, walking remains a sensible and, all things considered, safe choice.

The bottom line: from a danger-to-your-kid, risk-assessment perspective, there is no reason not to let your child walk to school. Of course, parents need to consider factors such as distance, age, maturity, and the nature of the traffic in their neighborhood. But an *independent active commute* should be the default. Also, we need to consider the consequences to the community of not walking. For example, as fewer kids walk, the less pressure there is to create safe streets. A 2014 study examined 10 years of data on pedestrian accidents associated with children walking to school. As with similar research, the study found that accidents were relatively rare and, as such, concluded that physicians should "counsel parents to encourage children to walk to school as a healthy lifestyle choice." In addition, it found that the most significant variable associated with accidents is the built environment, especially road crossings. This is good news: yet another study showing that walking is safe and can be made safer with safer crossings. But achieving that would take parental pressure, which won't be there if their kids don't walk to school.

Having more kids walk could also help to break the stranger-danger cycle of fear. A 2015 study from Australia explored the ways in which fear of strangers influenced parents' decisions to let their children walk. This study also found that the built environment was a key variable. Specifically, it found that the more walkable the neighborhood, the greater the number of kids walking—and the less fear parents had of strangers. For parents, eyes on the street matter.

We must throw into the calculus the profound and well-documented benefits of walking to school. Of course, exercise is a big one. Most children in North America don't get enough. And most parents don't realize the degree to which this is the case. In Canada, 88 percent of parents believe their kids get enough exercise, but in reality, only 7 percent do. Getting more kids to walk (or bike) to school won't solve this profound problem, but it would certainly be a step in the right direction.

In addition to the obvious benefits of increased exercise, a 2018 study from Sweden found that going to school without parents in tow allows kids to better develop their social skills. As noted by the study's author, when kids are driven to school they "lose natural opportunities to explore their neighborhood and to interact with friends on their own. As a result, they become less independent and secure in their immediate environment."

This study also found that children who walk to school performed better academically—a finding that is consistent with other research. A 2012 study from Denmark, for example, involving almost 20,000 students, revealed a connection between walking or biking to school and concentration—an effect that lasted the entire morning. It can have mental health benefits too, including reducing stress and anxiety.

Paradoxically, letting kids walk or bike on their own may, in the long run, also make them better at handling traffic. A Spanish study of almost 800 children between the ages of 6 and 12 found that those who actively commuted without a parent were more aware of safety issues than those who were accompanied. In other words, by driving or accompanying a child to school, parents are having an *adverse* effect on their kids' safety skills.

Given these realities, it is no wonder that some jurisdictions are implementing policies to encourage parents to forgo driving and to give their children more independence. In the U.K., some school districts fine parents for idling in front of a school. More dramatic is the 2018 law in Utah that aims to shield from liability parents who let their kids play independently. (Lenore Skenazy was a driving force behind it.) The legislation, which has been called the free-range parenting law, was designed to encourage things like letting kids walk to school alone. While all agree that the safety and the best interest of children should remain the primary considerations, these kinds of policies represent the recalibration of how we calculate what is best for our kids.

Risks do permeate today's world. But we shouldn't allow risks (and let's not forget that now is the safest time in human history) to be magnified in a way that warps our decision-making. To paraphrase Lenore Skenazy, let's not let the leveraging of fear, dread, and worry win the day.

In the end, a combination of irrational fears, cognitive biases, media hyperbole, marketing spin, and social-media-facilitated peer pressure have worked together to create a seemingly unstoppable momentum around the idea that driving is a reasonable option.

Resist.

This is what parents need to consider: On the active-commute side of the equation, there are tangible health, social, psychological, environmental, and educational benefits. On the driving side of the equation, largely unsubstantiated and certainly exaggerated concerns about strangers and crime and little or no evidence to support them. The choice—when feasible—is obvious: relax, dammit, and let your kid enjoy the walk.

7:50 AM—Getting to work

I suspect you know where this one is going, so let's start with the conclusion: if it is reasonably possible, embrace active commuting. The benefits are clear and well documented. One study found that people who cycle to work had a 30 percent reduction in all-cause mortality. A 2017 U.K. study of more than 250,000 participants found that, after controlling for a wide range of variables like wealth and education, cycling to work was associated with lower rates of cancer and heart disease. A 2018 study by a team at Cambridge came to a similar conclusion, finding that active commuters had a "30 percent reduced risk of dying from heart disease and stroke" and lower all-cause mortality. Studies have also found that cycling or walking to work can offer a significant emotional payback. A study of almost 18,000 commuters found that cycling to work had a significant positive effect on psychological well-being. It can lower stress, improve concentration, and elevate your mood.

Admittedly, this kind of thing is hard to study well, and much of this research is correlational in nature (for example, people who embrace active commuting might simply have rosier dispositions). Still, a large and growing body of evidence suggests that how we commute to work can affect our health and happiness. How much does commuting matter? A 2018 media study from the U.K. found that people would rank a short, pleasant commute above sex. (And unlike sex, the shorter the better.)

I appreciate that a soul-crushing, time-consuming, vehicle-involving, unsexy commute is often not a choice. Distance, or the need to have a vehicle for work or family activities, can make walking or cycling highly impractical. In the U.S., 76.3 percent of the population commutes to work alone in a car, while fewer than 4 percent actively commute. In sunny, warm San Diego, less than 1 percent opt to cycle or walk. Compare that to cold, dark, rainy Copenhagen, where around 40 percent of the population rides a bike to work. Clearly, we can do better.

Of course, many factors beyond distance are relevant to the decision not to bike or walk, and many of these are largely out of our control—such as the built environment (cities that are more conducive to cycling, like Copenhagen, make cycling easier, and that can help to create a cultural shift in how people think about transportation). Yet many of us make a deliberate decision to drive—despite the reality that another mode might be the healthiest and most efficient option.

What drives people to drive? Fear, fashion, and time.

Research has found that many people view cycling as an intrinsically unsafe mode of transport. In part, this is probably because cycling accidents often attract significant media attention—which may cause people to think the chance of a serious injury is higher than it actually is. (Once again, the availability bias at play.) An analysis of a decade of media coverage in the U.K. found that coverage of cyclist fatalities rose 13-fold from 1992 to 2012, far out of proportion to the coverage of, for example, accidents involving motorcyclists. The authors of the study believe this degree of media coverage creates a negative feedback loop that hurts interest in cycling. Indeed, a 2014 study, also from the U.K., found that 64 percent of the population believed it was too unsafe to cycle on the road. Only 19 percent disagreed.

Is cycling really this dangerous? The risk of death—the ultimate risk—is about equal to walking or driving. One analysis, published in the *Financial Times Magazine* in 2016, found that people are more likely to die walking in a city than cycling. Yes, cycling clearly has risks. Crashes occur. But the risks are not out of proportion to other transportation choices (except motorcycles—which are crazy dangerous). Statistics generated by the University of British Columbia break it down another way: there is one car driver or passenger death per 10,417,000 person-trips, one pedestrian death per 6,803,000 person-trips, and one bicyclist death per 7,246,000 person-trips. Again, driving, walking, and cycling, from a mortality perspective, carry relatively similar risks, especially when you consider the remoteness of the overall mortality incidence. Behind the wheel, there is, approximately, a 0.0000001 chance of death. When you hop on a bike, there is a 0.00000014 chance.

But what is key to making a decision about commuting is not the raw odds of something going wrong while you are on your bike. All forms of transportation—even walking—carry some risk. If you were to stay home and lie on your bed, covered in pillows, the chance of a transportation accident would be pretty low. But that isn't a good, long-term life strategy. The question, then, becomes: What is the best way, all things considered, to get somewhere? What is needed, at least in terms of health, is an analysis of the risks versus the benefits. Do the well-known and well-documented health and psychological benefits of active commuting (to say nothing of the environmental perks) outweigh the risks associated with accidents?

The answer is a clear and unequivocal yes. I could not find a single study, research review, or policy statement that suggested otherwise. For example, a 2015 systematic review of all the relevant data concluded that the "physical activity benefits exceed traffic-associated collision emission detriments." A 2010 analysis concluded that the health benefits of cycling "were substantially larger than the risks."

So, while we do need to make cycling safer—including changing the built environment of our cities—it is already almost always a healthy and logical choice.

Another fear that keeps commuters off bikes: they think people will hate them. Studies have shown that many of us believe the general public has a very negative view of cyclists. This means, basically, that people don't ride their bikes because they are afraid of offending drivers. Alas, they are kind of right. A number of studies have confirmed that drivers do think cyclists are annoying as hell. This distaste not only increases the chance that drivers will act aggressively toward cyclists, but also decreases the chance that the driver will actually *see* cyclists on the road—thus increasing the tense relationships between cyclists and cars and increasing the perception that cyclists are annoying!

From my own experience, this concern is completely understandable. I've been commuting to work on a bike for more than 25 years. During that time, I have been told to get off the f—ing road hundreds of times. Often, the driver's eyes are filled with an intense, inexplicable rage. It is as if I've committed a morally reprehensible act that has offended the driver to the core of his or her being. On one occasion I was riding down a bike lane on a quiet side street near the university where I work. A driver went out of his way to stop me to tell me to get on the f—ing sidewalk. I hadn't slowed his progress, cut him off, or run a stop sign. He just didn't like me being on the road. At the moment that this exchange occurred, my bike and I were stopped directly on top of a big, fluorescent bike emblem painted on the road to delineate the parameters of the bike lane. Fearful this guy was going to hop out of his car and beat me to death with my hipster single speed, I simply pointed at the emblem. He screamed "F— you!" gave me the double-middle-finger fist pump, one more "F— you," and sped off.

This hatred of cyclists is driven by a number of factors, and near the top of the list is the perception that cyclists don't follow the rules of the road. People seem to find this intensely aggravating, even though transgressions rarely impede the path of the driver. Moreover, research reveals that most cyclists *do* follow the rules—more so than drivers. A 2017 study by the University of South Florida's Center for Urban Transportation Research, for example, used cameras, GPS, and proximity sensors to track cycling behavior. It found that "the proportion of compliance with general traffic rules for bicyclists was 88.1%," which was slightly higher than it was for drivers (85.8 percent). A study from Australia used a hidden camera to monitor an intersection for six months. In total, they were able to assess 4,225 cyclists and found that only 6.9 percent broke the rules, mostly by turning left against a red (the equivalent of turning right in North America) without stopping and when there were no vehicles going in the same direction. In addition, a 2017 study from the University of Colorado found that when cyclists do break the rules, the dominant reason (70 percent) is to increase personal safety, whereas the main rationalization for drivers' rule-breaking was to save time (77 percent). When accidents do happen, they are almost always the fault of drivers. One study found that drivers are solely to blame 60 to 70 percent of the time, whereas cyclists are to blame 17 to 20 percent of the time.

I'm sure many readers will find these statistics nearly impossible to believe. Every driver has had to deal with an idiotic curb-jumping, havoc-wreaking, daredevil cyclist during rush hour. Indeed, a 2018 study from Canada found that, despite the empirical research to the contrary, most people blame cyclists for the road tension and cycling-related accidents. What is likely happening is that our cognitive biases take over. We vividly remember the few idiot cyclists—especially if we have a preconceived aversion that welcomes confirmation—but don't remember or even notice all the other cyclists. I also think that many people view cyclists as free riders (the road is for cars!) who are abusing their two-wheel mobility. Tom Stafford, a cognitive science researcher at the University of Sheffield, has suggested that we may be hardwired to be irritated by people who appear—rightly or not—to be breaking the moral code of traffic flow. Humans are social animals, and driving is a social act that demands cooperation. So when a cyclist in the bike lane zips past a traffic jam, the unconscious response is often rage, not "Gosh, that cyclist sure is lucky and smart!"

Regardless, should we continue to let the ill-informed, gut-reaction anger of unknown others dictate our commuting behavior? Nah. Haters gonna hate.

This brings us to the fashion deterrent. In North America, cycling is framed as an exercise. Commuters put on special cycling clothes and many ride to work as if participating in a competitive sport. As a result, when people think of bike commuters, they often conjure up images of dorky middle-aged riders in their distinctive—and often less than cool—bike commuting gear or of hard-core Tour de France wannabes.

This framing is bad. It creates a belief that you need to bring extra clothes with you to work and shower when you arrive. And if you hate the look of cycling clothes—and, let's be honest, only a tiny percentage of the population looks reasonable in head-to-toe Lycra—the thought of becoming part of the cycling horde can be less than appealing.

This framing also makes people think you need a degree of athletic ability to be a "cyclist." This perception is also wrong. For most commuting, you don't need special clothes, and you don't need to shower and change. Just hop on your bike and go. And remember, it isn't a race. Don't let those hard-charging, gear-smashing commuters turn you off the idea. Take a look at pictures of people commuting in Denmark, Belgium, or the Netherlands. There are multitudes of bikers wearing everything from jeans to suits to skirts and high heels.

I once spent some time as a visiting professor in beautiful Leuven, Belgium. My colleagues, young and old, would show up to work, dinners, and conferences on their bikes wearing whatever was required for the day. Once, I was meeting a senior professor for coffee downtown. She rolled up on a bike looking energized and amazing in her business attire, her briefcase and lunch in a basket on the front of her bike.

"How far away is your home?" I asked, expecting her to say she lived just around the corner.

"Oh, not far," she responded. "About five kilometers."

We are all in a rush. So it is no surprise that time and distance are found to be significant barriers to active travel. However, once again, perception often doesn't fit reality.

A 2018 study from Pennsylvania State University asked hundreds of people how long it would take to walk or bike to a variety of locations. They were almost always wrong, usually thinking it would take longer than it actually would. Fully 93 percent incorrectly estimated how long it would take to bike to a particular spot. As Professor Melissa Bopp, one of the authors of the paper, summarized: "Traveling by foot or bike has a lot of benefits, but not a lot of people do it. They may think they can't do it because it's too far and it'll take too long, when it turns out, it's really not."

The average Canadian commuter spends 25 to 30 minutes driving to work one way (which, incidentally, is among the highest in the world). And the median commute is a distance of 7.7 km—a distance a person can easily cycle, going at a modest pace, in under 30 minutes. So, for over half of all car commuters, driving does not save time, especially if you add the time required to park and walk to the door of your destination. And in places where traffic is bad (commuting almost always happens during rush hour), cycling likely saves commuting time.

I used Google Maps to get an approximation of how long it would take to bike 7 km or so in a range of major cities. Harvard Law School to Boston Commons is about 7 km and, according to Google Maps, that is a 31-minute bike ride. In Toronto, it takes 31 minutes to travel the 7.8 km between the CN Tower and Eglinton Avenue. Battery Park to Times Square in New York City is 7.9 km and would take 28 minutes. And in my home city of Edmonton—which is a prairie town, so pretty darn flat—you can travel a vast distance in 30 minutes.

I live 6.5 km from work, so my commute is a bit less than average. Riding at a relaxing pace, it takes me about 20 minutes in the summer and about 25 minutes in the winter. I absolutely love that time. I'm lucky, though. Most people hate their commute. Really, really hate it. In a 2014 survey of more than 900 women in Texas, conducted by Nobel laureate Daniel Kahneman, respondents said that the morning commute was the worst thing they did all day. But not everyone feels the same way. A Statistics Canada survey done in 2005 found that about 3 percent of us said that commuting to work was the best part of our day. Who was this strange group? It was dominated by cyclists. In fact, 19 percent of people who rode bikes to work said that their commute was the "most pleasant activity of their day."

Active commuting is not for everyone. You might be one of the few individuals who enjoy driving. Or perhaps you live a great distance from your workplace, or you need a car for work or after-work activities. Or the roads between your home and destination may simply be too treacherous for a bike. Fair enough. But don't let misinformation or misperceptions sway your decision. An active commute is better for health and happiness. It is better for the environment, and it helps reduce traffic congestion. It is often more time efficient and can save you money. For me, the choice is obvious.

Now, when I hear "Get off the f—ing road," I simply smile and wave. Heck, I'm not the one stuck in a car.

8:15 AM—Parking

Okay, let's say you drive your car anyway. You simply must. Once you get to your destination, especially if you live in a big city, you will now face one of the most irritating aspects of human experience: parking.

A 2017 survey from the U.K. suggests that the majority of the population regularly feels stressed about finding a parking spot. The study found that two-thirds have canceled a family outing because they couldn't face the stress of parking, and 40 percent have missed an appointment. In the U.S., one-third of male drivers say they have gotten into a confrontation with another driver over parking in the previous year.

It is no wonder people feel this way. Parking consumes a ridiculous amount of time and psychic energy. It is an activity that is both tedious and anxiety-provoking, like dishwashing at gunpoint. If you add up all the time spent searching for a spot, it amounts to dozens of wasted hours every year. In New York City, for example, the average car owner spends a whopping 107 hours a year (that's 4.5 days!) searching for parking. An economic analysis from 2017 estimated that all this wasted time in search of parking costs the U.S. economy over $71 billion a year.

Humans behave horribly in parking situations. A 1997 study found that when someone is waiting for a parking spot, the driver in the parking spot takes longer to leave than if there was no one waiting. The researchers called this "territorial defense," suggesting we have a natural tendency to retaliate against waiting drivers—even if there is no reason to do this and, in fact, the vengeful slowness is almost always against our own interest in saving time.

Even though we all seem to be hardwired to be parking lot jerks, we nevertheless expect others to behave civilly. And we are enraged when they don't follow the unspoken parking lot code of ethics. My son

Michael and I once went to a big-box store near closing time. The large parking lot was almost empty. We were in a hurry and I did a crap job parking. When we got out, I saw that our VW was slightly angled across one of the parking lines.

"I'd better move the car," I told Michael.

"Come on, the parking lot is totally empty," he said, heading toward the store.

"I bet we have a note on the car when we get back," I predicted.

Twenty minutes later, we emerged from the store to find a note on the windshield. It read: "The people of the world want you to learn how to park. You can do it. It isn't that hard. Signed, everyone."

While much of our parking stress is out of our control (there simply aren't enough parking spaces in many cities), our parking decisions have consequences. Indeed, the simple act of parking stands as a wonderful example of how our cognitive biases and ill-informed habits result in frequent bad decisions that can make our day just a little bit worse and, over time and as the bad decisions accumulate, can have significant social costs.

People spend a lot of time (read: waste a lot of time) searching for the perfect spot. We are compelled to do this, at least in part, by the memory of the last time we had a good spot. We strive to replicate that pleasurable experience. *Seinfeld*'s George Costanza went so far as to put finding a good parking spot in the same category as sex. "Parking in this city is just like sex," George said. "Why should I pay for it when, if I apply myself, I might be able to get it for free?"

Andrew Velkey, a psychology professor at Christopher Newport University in Virginia, agrees with the last-great-spot theory and the thrust, so to speak, of Costanza's categorization. Professor Velkey, who has studied parking lot behavior by, well, watching a lot of parking lot behavior, told me that people more vividly remember those rare instances "when they secure a 'great parking' spot." As a result, and like George Costanza, they feel compelled to apply themselves every time they are faced with a tough parking situation. "What they don't remember as well are the more likely instances when they secure more typical spots," Velkey continued. "As a result, people over-estimate the likelihood they will secure great spots."

Snagging a good parking spot feels like an accomplishment, as if you've scored a goal in the ferocious sport of car parking. As Velkey noted, "There is a sense of accomplishment, a sense of satisfaction, maybe a little bit of victory because you know you're competing against other people—'I got the space, you didn't.'"

Velkey has also found that people simply do not learn from their past parking experiences. "The more experienced car parkers made the same mistakes that less experienced drivers make," he told me. "People don't approach parking as an opportunity for deliberate practice. As a result, they don't improve much."

In summary: you aren't some all-powerful parking savant. Don't apply yourself. This isn't a competition.

In fact, the best strategy is to ignore these compulsions. People who search for a good spot usually spend more time getting to their destination than those who simply take the first spot they see. A study published in 1998 used mathematical modeling to predict the best approach to finding a spot. It found that simply picking a row closest to the parking lot entrance (which is often the row farthest from your destination) and parking in the first available spot wins over a more aggressive strategy of looking for the space closest to your destination. The time you spend searching for a space usually negates any time you save from being closer. And that just adds aggravation to your day.

Although this kind of modeling is usually done with parking lots, the logic likely holds for parking on the street. Park in the first spot you find, even if it is a good distance away from your destination. Plus, why not enjoy the walk?

When it comes to parking, don't be a jerk, tolerate other drivers' jerkiness (remember, you probably come off as a jerk too), and park in the first spot you find open.

8:30 AM—Start work

Do the creative, hard, big important stuff first. Do not get sucked into an email vortex. Be strong.

Studies have consistently shown that we perform better at certain times of the day, and for most of us that is the morning. A University of Chicago study from 2016 analyzed data from almost two million students, 6th through the 11th grade, to get a sense of how much the time of day affected their performance. It found that students learn more in the morning than later in the school day. Taking math and English in the morning significantly increases students' GPAs.

9:30 AM—Public toilet seat

I'm a bit of a germophobe, so the toilet seat issue is personal. If hands-free bathrooming ever becomes an Olympic sport, I feel good about my chances of making the national team. Indeed, my lavatory angst has caused me to dig deep into the bathroom-stall germ-distribution research—yep, this data exists—in order to inform my day-to-day selection of toilet seats. What does this scatological scholarship reveal? If germs are your concern, the first stall is probably your best bet. Germophobes take note: *avoid the middle stall!*

I'm not the only person who is leery of the loo. A study from 1991 found that a remarkable 85 percent of women squat over public toilet seats. No touching. (Given that women spend substantially more time in the bathroom than men—yep, this data also exists—that is a lot of bladder-induced isometric strength training. Perhaps that is why women live longer?) An additional 12 percent of women cover the seat with paper, and only 2 percent sit on the actual porcelain. I'd classify that as a remarkably broad-based fear of the necessarium. Not only is this fear a hassle (it can be difficult to time biological breaks to perfectly coincide with the availability of a spotless facility), but it can also have health consequences. A 2017 Swedish study of 173 university students, for example, found that "irregular or infrequent voiding due to avoiding school toilets can contribute to a number of urinary problems."

Many factors contribute to the avoidance of public washrooms, but the core concern seems to be germs. But should we really be worried about germs on the toilet seat? Should you use one of those toilet seat covers to give yourself a layer of protection from all the nasty germs residing on the loo?

While public toilets can appear pretty grimy, it is unlikely that they pose any serious germ-related risks. Contrary to the many heartfelt and authoritative warnings I received from my junior high classmates, you can't catch a sexually transmitted infection from a toilet seat. In fact, the vast majority of microorganisms are pretty harmless to begin with, and most won't last long on bathroom surfaces. It would likely take an absolutely perfect-storm germ-transmission scenario for an infectious agent to pass—such as sitting on a toilet seat immediately after an infected person, with an open wound that seamlessly lined up with the pathogenic, or disease-causing, microbe. But even under such circumstances, the risks are still remote. Dr. William Schaffner, a professor of preventive medicine at Vanderbilt University, has concluded that "toilet seats are not a vehicle for the transmission of any infectious agents—you won't catch anything."

Professor Brett Finlay agrees. He is a colleague of mine, a renowned microbe and immunology researcher, and the author of the germ-embracing book *Let Them Eat Dirt.* "Toilet seats actually aren't that dirty," he told me. "Being hard plastic, they aren't so bad. Ironically, sinks are worse, as they are wet and form biofilms"—that is, colonies of microbes. The bottom line, Professor Finlay told me, is that "hard dry surfaces don't have many germs, soft wet surfaces do."

Studies have consistently found that many common items have far more germs on them than public toilet seats, including cellphones, television remotes, gym equipment, kitchen cutting boards, and the kitchen sink faucet handle. So you can sit on the toilet seat with some degree of confidence.

And those paper toilet seat covers are almost certainly useless. They are porous for one thing, so tiny bacteria and viruses can easily pass through the paper to the epidermis on your behind (not that that really matters, as noted above). Moreover, toilet seat covers do little to stop the truly potentially problematic aspect of the defecation process, the "toilet plume"—which is, as summarized in a 2018 study on the phenomenon, "the aerosolization of fecal matter during toilet flushing." This study carefully measured how far bacteria spread with each flush. The conclusion? Pretty darn far, particularly when the flush has some force—as is often the case with public toilets—and when smaller bacteria are involved. A 2013 review of toilet plume data concluded that "potentially infectious aerosols may be produced in substantial quantities during flushing" and that "aerosolization can continue through multiple flushes to expose subsequent toilet users."

That sounds pretty nasty. But, again, we should keep our fecal fears to a minimum. Yes, the rare harmful pathogen may be transmitted via aggressive toilet flushing, but from the perspective of individual risk, the potential for toilet-induced harm is pretty small. As Professor

Finlay reminded me, "Fecal-oral route is by far the most common way of picking up enteric pathogens." In other words, it isn't the toilet seat or the microbe-filled toilet plume that you should fear, but your own dirty hands.

Now that I've quelled your fears, let's take a moment to celebrate the toilet. In the context of disease prevention and quality of life, it is perhaps one of the greatest of all human inventions. In 2007, the readers of the *British Medical Journal* picked advances in sanitation—basically, clean water and toilets—as the single greatest medical advance. Not stem cells, genetics, anesthetics, transplantation, or even antibiotics, but sanitation. Think about that the next time nature calls. You are sitting on a device that has saved millions and millions of lives. Unfortunately, much of the world's population still does not have access to toilets. In 2017 the World Health Organization reported that almost 900 million people must defecate in the open. And, because of population growth, that number is likely to increase.

A final toilet seat side note: One aspect of the bathroom experience that some don't need to worry about as much as others? Spit. Many men (as in every single guy that pulls up next to me at the urinals) have the ridiculous, disgusting, and seemingly primeval habit of spitting before they pee. As a result, in addition to urine splatter, toilet seats and urinals in public restrooms are often covered with gobs of spittle. What the hell is going on here?

Although I could find little empirical work on this point, some academics have speculated that it serves as a way to mark territory. (Why someone would want to mark a toilet at a sports stadium, bar, or movie theater as part of their territory is beyond me.) Men do not produce more spit than women. There is rarely a true physiological reason to spit. So please, men of the world, either improve your accuracy or find a way to resist your primitive urge to stake a claim to public porcelain.

While the risk of getting a serious disease through spit is pretty low, there is a small concern that spitting, just like the toilet plume, can increase the chance of spreading airborne bugs such as norovirus (responsible for so-called stomach flu). Indeed, during a particularly bad flu season in the U.K., the Health Protection Agency tried to get professional soccer players to spit less. "If they are spitting near other people," an agency spokesperson said, "it could certainly increase the risk of passing on infections." I'm guessing that this public health plea did not slow the flow of phlegm from the footballers one iota.

9:33 AM—Wash hands

Wash. Your. Hands. For the love of God, wash your damn hands! I've seen far too many people in the bathroom fling open the stall door and march straight out of the facility, zipping and buckling and tucking as they go. If I had a taser . . . zap!

Handwashing isn't a decision that requires a great deal of thought. Just do it. Even if you believe you use toilet paper with surgical precision, the above-mentioned fecal-filled toilet plume can deposit pathogens on your body. You may think your hands are absolutely spotless, but harmful germs can still be present. A single gram of human feces can contain one trillion germs. This is why, as proclaimed by the CDC, washing your hands "is one of the most important steps we can take to avoid getting sick and spreading germs to others."

It may sound as though I'm walking back the anti-germophobia message from the preceding section. But while toilet seats are relatively benign, many pathogenic germs *are* transmitted by the hands—often because we can't stop ourselves from touching our mouths, eyes, and/or nose. In addition, there is lots of evidence that tells us handwashing works, especially when it is done well. A 2014 study of elementary schoolchildren found that teaching kids how to wash their hands resulted in 36 percent lower absenteeism from stomach flu. Effective handwashing can reduce the incidence of everything from the common cold to more serious and life-threatening diseases. This isn't a trivial issue. On the global scale it has been estimated that handwashing could save millions of lives every year, primarily through a reduction in deaths associated with diarrhea.

Given the effectiveness of handwashing, it is frustrating that many people still do not do it—even though most of us think it's a good idea. A 2015 study from the U.S. found that 92 percent of Americans believe it is important to wash one's hands after a visit to the washroom, but only 66 percent said they actually do so. Worse, 70 percent admit to not using soap. When measured more objectively—that is, by counting how many people actually wash their hands—the numbers are even worse. And, no surprise, men wash their hands less often than women, and we do a far worse job. However, in general, all of us are pretty bad at this modest task. A 2013 study that observed 3,749 people and how they washed their hands found that only 5 percent did the job correctly.

People are far less likely to wash their hands when they're alone. A bit of peer pressure appears to nudge our hygiene habits in the right direction: a 2014 study by Stanford University used video surveillance to track handwashing at urban schools in Kenya. When someone else was present, 71 percent of the students washed their hands. However, only 48 percent did so when they were alone. Another study, done in 2009 by the London School of Hygiene and Tropical Medicine, found that when it comes to trying to persuade people to wash their hands properly, the most effective messaging invokes peer shame. Specifically, it found that a bathroom sign that said "Is the person next to you washing with soap?" was more effective than something with a nagging vibe, like "Don't be a dope—wash with soap!"

And what *is* the correct way to wash your hands? No, a quick rinse under the tap and a wipe on your pants doesn't cut it. You should aim to wash your hands with soap and water for 15 to 20 seconds. This is the time it takes to sing "Happy Birthday" twice. (Feel free to do the singing silently.) And wash both sides and between the fingers. You don't need to scrub old-school surgeon style, but put in some effort.

What about drying your hands? Fancy high-tech hot-air hand dryers seem to have invaded public bathrooms throughout the world. I suspect this move was motivated by cost (fewer paper towels to buy), environmental concerns (fewer paper towels to dispose of), and a desire to exploit the public health push to get people to wash their hands. The hand dryer industry is growing fast and is projected to be worth $1.6 billion by 2024. One industry report suggests the expansion is driven by "the increasing government initiatives to spread awareness about hygiene and cleanliness."

But these hand dryers aren't the best choice from a hygiene perspective. Research has consistently shown that paper towels are better. A 2012 review of the evidence concluded that paper towels are superior to electric air dryers and, as such, recommended the use of only paper towels "in locations where hygiene is paramount, such as hospitals and clinics." For a germophobe like me, this conclusion screams *Don't use electric hand dryers!*

And the hand dryer story only gets worse. A 2018 study found that electric hand dryers function as fecal particle distribution machines. They suck in the germy particles from the room—including the aforementioned toilet plume material—and blow them onto your hands. The researchers found that "many kinds of bacteria, including potential pathogens and spores, can be deposited on hands exposed to bathroom hand dryers." These machines also help to disperse germs throughout the entire building.

Okay, a bit of a risk-related reality check is needed here. The vast majority of germs flying out of hand dryers are perfectly harmless. Still, given that the objective of handwashing is to *remove* germs, the electric hand dryer system seems, at a minimum, counterproductive. So it is probably best to use paper towels.

When I told friends I was writing a handwashing section for this book, a lot of them told me that their preferred method of drying involved their pants. "Oh, I just wipe my hands on my pants. Doesn't everyone?" was a common response. I must admit, it's a strategy I frequently utilize, even when other options are available. Your pants are just right *there*, ready for the wiping.

Is this a good idea? I could not find any research that explored this on-the-fly drying tactic. But your pants can harbor bad germs, especially if you've pooped or vomited on them recently. (If that is your situation, you probably have bigger issues to deal with than a hand dryer/paper towel/pants decision.) That said, scientific investigations into the role of clothes in the transmission of infectious disease, such as in hospitals, have generally determined that there is a lack of high-quality evidence. It has been noted that it is certainly biologically plausible to transmit germs via clothing, hence the adoption of "bare below the elbows" policies by many hospitals around the world. Still, as long as your pants are excrement-free, you are probably safe to use them for a quick, efficient hand wipe. Skin microorganisms, which are generally what you'll find on your pants, are usually not hazardous to the source of those cells (i.e. you).

In 2011 a University of Alberta professor worked with a student on an interesting germy jeans experiment. Professor Rachel McQueen had one of her students wear one pair of jeans every day for 15 months straight, without a single wash. (Curious if the research ethics board had any concerns!) It turned out that the unwashed jeans had the same amount of germs as a pair that had gone just a few weeks without a wash. And those germs were not particularly nasty, just more of the usual mostly harmless skin microorganisms.

I'm not advocating that you stop washing your jeans, though this is now a thing among raw denim aficionados. But if you wash your hands regularly with soap and water, there is no evidence that the occasional jean-swipe hand dry is going to present a problem. That said, your dirty jeans can still have nasty germs, especially if you go commando (remember "Get dressed" above?), simply because, as Professor McQueen put it, "pants are in close contact with one's anogenital region."

10:00 AM—Another coffee?

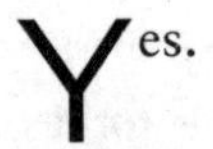es.

10:30 AM—Multitask

I was once a guest on a local radio show where the topic of discussion was multitasking. Should we do it? *Can* we do it? The person who spoke before me was a human resources expert. It is her job to find people employment, and she didn't pull any punches. "The number one skill employers are looking for right now is the ability to multitask," she told our listeners. "Employers expect people to do many things at once. This is the world we live in!"

Alas, she is right. One study found that the average office worker is interrupted once every three minutes and five seconds! And about half of those interruptions are self-inflicted. Most people fail to focus on one task for more than three minutes. We check our email. We click on a pop-up news alert. We scan our Twitter feed. We glance at Kim Kardashian's Instagram account. We are a society of ridiculously distracted people!

There's a lot of interesting research on workplace distraction and multitasking. Studies suggest that people spend—if you include both your laptop and your mobile phone—anywhere between five and eight hours a day online. A large percentage of the population spends more time on phones and computers than sleeping. Increasingly, the world is engineered for distraction. It's engineered for multitasking. Market forces are at play. News alerts sneak onto our computer screens. Pings arrive with each new text message and Facebook update. And it is ridiculously easy to get pulled into the vortex of online surfing. You have an idea, you go looking for something relevant to it, and before you know it you find yourself hopping from website to website.

I'm a writer. I'll often sit down and think, "I'll give myself half an hour without checking my email, or my Twitter account, or surfing." And it's hard. I find once I get that groove, the hours will pass. But until you find that groove, it's really difficult, especially if you are pulled online for the purpose of research.

Constant distraction takes a toll. Studies consistently confirm that multitasking and interruptions lessen our performance. Our brains can process only so much information at once, and so multitasking creates a cognitive drain: the more you switch between tasks, the less efficient you become. Once you are pulled away from a task, it can take a significant amount of time to reengage. One study found that it takes more than 23 minutes, on average, to get back on task. As well, multitasking increases worker stress, costs the economy over $650 billion in the U.S. alone, and, according to a 2015 survey, causes people to waste, on average, about three hours a day.

All told, there is very little evidence to suggest that the decision to multitask is a good one.

A bit of research suggests that a very small percentage of the population, about 2.5 percent, can effectively multitask. Unlike the other 97.5 percent of the population—that is, nearly everyone—these "super-taskers" don't experience a significant performance reduction when trying to balance two things at once, at least in controlled experiments. But if you think you are one of these rare individuals, you almost certainly aren't. A 2013 study from the University of Utah found that people who think they're good at multitasking are often particularly bad at it. They are distracted more quickly and they can't concentrate, which is probably why they are multitasking in the first place. As the lead author, Professor David Sanbonmatsu, was quoted as saying, "People don't multitask because they're good at it. They do it because they are more distracted." The tendency to multitask was "negatively correlated with actual multi-tasking ability." And in a classic example of the we-all-think-we-are-better-than-average phenomenon, 70 percent of the undergraduates in the study thought they were better than average at multitasking—which, of course, is statistically impossible.

Another myth of the digital age is that young people (the archetypal "kids these days") are better at multitasking and handling a constant stream of information simply because they were born into this reality. But the idea of a "digital native" is wrong. A 2017 review of the evidence found that there was "no such thing as a digital native who is information-skilled simply because (s)he has never known a world that was not digital." The younger generation multitasks like the older generation: poorly.

Finding focus has never been more difficult. I actually play games with myself: if I finish this task, I'll buy myself an espresso. It sounds simple, but it works. Finish one task, then move on to the next.

11:00 AM—Anticipate lunch

You've had a productive morning. You got up nice and early, because, dammit, you are a morning person. You had a healthy raw-milk-free breakfast. You got the kids off to school and commuted on your bike. At work, you tackled your big, hard projects first, ignoring the mountain of email, news alerts, and social media pings. Lunch is just an hour away.

The idea of a looming appointment, even for a pleasurable event, can interfere with productivity. Bounded intervals of time—such as a block of time before a scheduled appointment—feel shorter than an unbounded interval of the same amount of time. More interesting, people are less productive when an appointment looms. In other words, structuring your day can actually make you *less* efficient. Professor Selin Malkoc of Ohio State University has speculated that this happens because we begin to direct our attention to the appointment. We start to mentally prepare for the event, making it feel like you have less time than you really do. Malkoc and her colleagues found that merely thinking about a future appointment causes us to do less work.

How can you avoid this schedule-induced downturn in efficiency? In addition to simply reminding yourself that time is not accelerating, try to schedule appointments and tasks close together. This minimizes the hunks of your workday that can be affected by this phenomenon. In addition, a tight schedule can make room for more unbounded and productive work time.

II

AFTERNOON

12:00 PM—Lunch

I was in New York City for work once, and after a busy morning I was desperate for an overpriced double espresso and something sweet, preferably a huge chocolate chip cookie. (Yes, I realize this is a spectacularly poor lunch decision. So it goes.) I used my smartphone to locate a few cool-looking cafés. The first one I found had the requisite NYC atmosphere: vintage lighting, hip minimalist furniture, indifferent staff, and an epic-looking espresso machine. But what it didn't have was gluten. The baked goods *looked* wonderful, but they were all gluten-free. I wasn't in the mood to forgo gluten. On to the next café.

This establishment also had the desired you-are-gonna-pay-a-lot-for-our-coffee interior design. But, again, it was almost entirely gluten-free. Desperate, I settled on the one full-gluten offering, a plain croissant. While paying I asked the barista why there was no gluten.

"It's healthier," she said flatly, without looking up from her barista activities. She clearly thought the answer was obvious.

I was about to launch into a lecture about how stunningly misguided this statement was and how this belief is, in fact, part of a broader social trend, fueled by marketing pressure and a constant stream of pop culture misrepresentations, that was undermining the public's ability to make healthy food decisions. But I sensed that hearing my food policy philosophy was not a priority for this particular barista.

"Right," I replied.

Thankfully, the croissant was fabulous.

In many ways, deciding what to eat can be one of the most confusing decisions we make all day. And it is becoming more and more confusing. Every day there seems to be contradictory news about how we are supposed to eat. And every few months a new, fully formed dietary fad appears out of the ether encouraging us to eat more or less fat, more or less carbohydrates, or more or less protein. We are to fast or graze or juice. We should eat superfoods or just this or that food. And we are to avoid sugar, high glycemic-ish stuff, and, of course, gluten.

Indeed, the speed at which popular thinking evolves makes it difficult to write a book that critiques nutrition and diet fads. By the time this book comes out we might be in the grasp of an all-sugar-all-the-time diet craze. (This may sound like an absurd stretch, but just think about our shift in attitude about fat!)

The gluten-free trend is an example of how our decision-making around food can go seriously off the rails and, unfortunately, stay there. I first wrote about this topic in 2013, when it was already a very popular celebrity-fueled fad. It was a short piece in the *Toronto Star* questioning nutrition advice provided by the singer Miley Cyrus that claimed going gluten-free would help you lose weight and be healthier. At the time I thought the fad would be brief. But here we are, years later, and I can't find a gluten-filled chocolate chip cookie anywhere. How did we get to this wheatless state?

If you have celiac disease—a serious autoimmune disorder that affects about 1 percent of the population—you must go gluten-free. And a few percentages more of us may have non-celiac gluten sensitivity, a diagnosis that is somewhat controversial, although more research is emerging. But if you believe the marketing data, almost a full third of us are trying to go gluten-free. A 2017 industry analysis found that 27 percent of the U.S. population had purchased gluten-free products in the previous three months. According to Agriculture and Agri-Food Canada, about 10 million Canadians are trying to go gluten-free. Those are huge numbers.

More interesting is *why* people are going gluten-free. The 2017 study from the U.S. found that the top three reasons, by far, are: to try something new, because it is healthier, and to lose weight. Other research has found that the vast majority of people who avoid gluten have no symptoms of intolerance or sensitivity. And a 2018 study of young adults found that many valued gluten-free food because of its purported health qualities. (They tended to be the same kind of people who also valued organic, non-GMO, and unprocessed foods.) Another study found that people view gluten-free foods as "more healthful, as having fewer calories, and as less processed" compared with foods containing gluten. This suggests that gluten-free has taken on a "health halo"—that is, it is viewed as being inherently healthy, even if it isn't.

The degree to which this belief about gluten has been absorbed by our culture is also illustrated by a 2015 study of more than 900 elite non-celiac athletes. In this cohort, one that included world champions and Olympic medalists, a remarkable 41 percent had adopted a gluten-free diet because they believed it was healthier and would improve their performance.

Here is the reality: There is absolutely no evidence to support the idea that going gluten-free is a healthier way of eating. And there never has been. This is also true of the claim that it is a good way to lose weight. In fact, the public perception that gluten-free food is healthier is entirely incorrect. It can even be harmful. A 2017 cohort study of over 100,000 people, for example, found that "the avoidance of gluten may result in reduced consumption of beneficial whole grains, which may affect cardiovascular risk." As a result, the authors conclude, "the promotion of gluten-free diets among people without celiac disease should not be encouraged." And avoiding whole grains may also increase your risk of colorectal and other cancers. Unless you have a clinically diagnosed reason to go gluten-free, it is, in the aggregate, a less healthy option.

Moreover, gluten-free products often contain unhealthy ingredients. To cite just one example, a 2018 study published in the journal *Pediatrics* examined the nutritional quality of gluten-free products aimed at kids. What the research uncovered was that these products were "not nutritionally superior to regular child-targeted foods and may be of greater potential concern because of their sugar content." The author of this study, Professor Charlene Elliott, concluded that "the health halo often attributed to the GF [gluten-free] label is not warranted, and parents who substitute GF products for their product equivalents (assuming GF products to be healthier) are mistaken."

And let's not forget that gluten-free foods are often much more expensive than their gluten-containing counterparts. One study found that, on average, gluten-free products were 242 percent more expensive than regular products.

And in case you are wondering, the evidence does not support the idea that going gluten-free will enhance the athletic performance of a non-celiac.

Finally, there is no evidence to support the weight loss benefits. On the contrary, evidence suggests a gluten-free diet can lead to weight *gain*. A 2019 study from the University of Chicago found a rise in the BMI of adolescents on a gluten-free diet. The authors speculate that this is due, at least in part, to a "surge in production of processed gluten-free foods on the market."

Despite all this science suggesting no benefit, the market for gluten-free products continues to grow. While gluten-free doesn't dominate public discourse as much as it did a few years ago—the current darlings are the ketogenic diet and intermittent fasting—it remains a big business. The global gluten-free food market is predicted to rise from $4.26 billion in 2016 to $7.38 billion in 2021. In an effort to capitalize on the trend's health halo, new gluten-free products are constantly being introduced, including gluten-free pet food, shampoo, and water. (Yes, someone is selling gluten-free water. The website for the product claims it is "gluten-free, non-GMO, certified kosher, halal, and organic." It's just water. H_2O.)

Professor Elliott told me, "I think the gluten-free trend, like the fat-free trend before it, captures attention because it promises an easy fix to consumers' nutritional and health worries." Elliott is an expert in food marketing and a Canada Research Chair at the University of Calgary. She noted that reliable studies are often drowned out by marketing and promotional claims. "The best-selling diet books and the testimonials on social media insist that gluten-free is the path to weight loss, health, and boundless energy. Health halos have huge cultural traction, and are difficult to unsettle," Professor Elliott said.

This is a remarkable situation. Gluten-free as a diet is clearly not healthier, not a good weight loss strategy, does not improve athletic performance, is more expensive, and, unless you have a clinically diagnosed reason to go without gluten, may even be harmful. Despite all that, many people view it as a healthy choice, and because of this perception it has become a massive industry.

(It is worth noting that for people with celiac disease, the gluten-free trend has mostly been a boon, because more food options are available. However, a study from 2018 found that the growth of the gluten-free industry is actually a "double-edged sword" for this community, as "they are increasingly faced with misunderstandings about the severity of celiac disease as a result of many nonceliac disease individuals subscribing to the gluten-free diet.")

For our purposes, the story of the gluten-free diet is a good example of how a pop culture trend can morph into a broad public misperception that, in turn, is adopted by the market to sell products that further entrench the science-free misperception in the public consciousness. Once the health halo takes hold, it is difficult to dislodge.

Alan Levinovitz is a professor at James Madison University and author of *The Gluten Lie: And Other Myths About What You Eat*. I asked him why the gluten-free health halo is so enduring. "Humans have a tendency to generalize from specific cases," he speculated, "and since gluten is a serious problem for some people, it's easy to believe that it must be intrinsically evil. Food is especially subject to this kind of generalization because eating is an extremely personal ritual that helps define personal identity and community membership—and because being healthy really does depend, in part, on eating well."

Professor Levinovitz touches on a key point about personal identity and community membership. Following a gluten-free diet has become equated with virtuous, healthy eating (GMO-free, organic, locally grown, and gluten-free) and is increasingly viewed as a form of self-expression. Of course, this is exactly why hip New York City coffee shops have embraced the trend. It fits with the gestalt they are trying to project. And once something becomes part of people's personal identity—as we saw with raw milk—it can be very difficult to change minds.

More broadly, the gluten-free trend is a good reminder not to let your food decisions be influenced by nutrition "noise." I recognize that food decisions are fantastically complex. And nowadays it is even more difficult to separate good advice from bunk—especially since it seems as if the nutrition research community is always changing its mind about what is healthy and what is not. The public finds this shifting advice confusing and frustrating. A 2017 study found that almost 80 percent of consumers say they regularly encounter conflicting nutrition advice, and 56 percent said that this conflicting data causes them to doubt their food choices. Indeed, the perception of scientific uncertainty may also cause people to trust science less. The perception that there is great uncertainty about nutrition science makes room for nutrition fads like the gluten-free one.

The good news is that, for the most part, when it comes to the nutrition machinations circulating in popular culture, you can ignore almost everything. This may sound like a ridiculous overstatement, but really, there is no magical diet. Can you name a single nutrition trend that, over the long term, has held up? In fact, although it may seem there is constant flip-flopping about nutrition advice, the basics regarding a healthy diet have been known for a long time and haven't really changed. Eat fruits and vegetables, whole grains, and healthy proteins. Aim for whole foods. Remember that there are no magical superfoods, and don't consume too many processed products. Basically, eat real food—and not too much of it.

1:15 PM—Rant!

Ranting is deeply ingrained in our culture. It is portrayed as cathartic! An emotional release! Let it out! You shouldn't bottle things up or they will fester and become a bigger issue!

Some of the most famous movie scenes are built around the dramatic punch of a good rant. "I'm as mad as hell and I'm not going to take this anymore" are the 13 words, delivered in an Oscar-worthy holler by Peter Finch, remembered from the 1976 movie *Network*. Jack Nicholson's "You can't handle the truth" is, for me, the only thing that remains of 1992's *A Few Good Men*. (I also have a vague memory of Demi Moore being skeptical of Tom Cruise's lawyering skills, but then not.) And it seems like virtually every single TV or movie psychologist, psychiatrist, therapist, marriage counselor, or understanding best friend has advised a struggling protagonist to face his/her anger and take a cathartic step toward emotional resolution (in other words, rant).

Ranting has probably been around as long as there have been annoying humanoids worthy of a rant. (I'm pretty sure that the "dawn of man" sequence at the start of Stanley Kubrick's *2001: A Space Odyssey* is really about the birth of ranting, which is later echoed by HAL the computer's murderous rant.) But social media has kicked ranting into the stratosphere, offering boundless opportunities. Now we can rant to thousands (or, if you are a celebrity, millions) and directly at the social media incarnations of the entities that infuriate us—airlines, insurance agencies, and the government (enough with the chemtrails!).

The internet has allowed us to vent in new and creative ways, using photos, memes, videos, and short, ungrammatical bursts of fury. Dr. Ryan Martin, a professor of psychology at the University of Wisconsin and an expert on anger management, reported that 46 percent of Twitter users admit that they use the platform to vent anger. And it seems we all love to watch the unfolding of high-profile rants. Indeed, research has consistently found that negative social media posts spread faster and further than positive ones. A 2014 study from China with the depressing title "Anger Is More Influential Than Joy" examined more than 70 million tweets and found that, yep, anger spreads faster than other emotions. It should be no surprise, then, that entire careers have been built around the public ranting formula. It helped to get President Trump elected, and social media ranting is a defining characteristic of his presidency.

So, when you get back to your office after lunch and find a group email from that annoying colleague—the one who likes to subtly imply you lack competence—do you respond with a rant? Or perhaps there is a particularly offensive missive in your social media feed. Is a good public rant in order? Will ranting make you feel better?

The answer is almost always the same: no.

Sigmund Freud deserves a big hunk of the credit for the catharsis myth. Core to his now almost completely discredited psychoanalytic belief system was the idea that getting our anger out was good for, or even essential to, our psychological well-being. For Freud, catharsis—which comes from the Greek *katharsis*, meaning "purification" or "cleansing"—was a way to relieve stress and unconscious conflicts. The cultural sway of Freud should not be underestimated. Although he did much to facilitate the secular exploration of the mind, his pronouncements were rarely, if ever, supported by empirical evidence. Nevertheless, his ideas continue to have tremendous cultural traction.

(Thanks to Freud, will a cigar ever be just a cigar?) Almost any theory that suggests that suppressed emotions can have an adverse effect on our physical and mental health flows directly or in part from Freud's writings on catharsis and the unconscious mind.

Many religions have also embraced catharsis to some degree. As Irish-heritage Catholics who never practiced, my family doesn't have much (okay, zero) experience in the confessional, but we've heard about it often. "Not sure you're ranting to God, but you are definitely letting things out," my brother Case mused to me during a discussion about this issue. Elements of forgiveness and repentance make such rituals more than mere ranting. But religious commentators have noted that practices like the confessional are linked with the idea of catharsis. A 2009 paper published in the *Canadian Family Physician*, authored by a physician and a priest, speculated on the therapeutic benefits of confession. "Even without delving into the theology of confession, its cathartic nature is evident. Uninterrupted, the penitent confesses until his or her list is exhausted."

From the perspective of therapy, probably the best-known, full-on catharsis-focused approach is the primal scream. This therapy became popular in the 1970s and involved screaming at the top of your lungs, outbursts of extreme emotions, and hitting stuff. It was made popular by psychologist Arthur Janov and his bestselling 1970 book, *The Primal Scream*. Many celebrities, among them John Lennon, were ardent followers of the method. Indeed, it has been suggested that Lennon's first solo album, *Plastic Ono Band*, was inspired by his experience with primal scream therapy. The album is peppered with unrestrained vocal shrieks. Lennon's screaming of the lines "Mama don't go" and "Daddy come home" near the end of the opening track, "Mother," certainly exudes a cathartic, primal-scream feel.

In more recent times, the idea of holding in your anger or emotions has been portrayed as a disease risk, something that could manifest itself in a tangible way, such as a tumor. Many alternative medicine practitioners have embraced the idea that bottled-up "toxic emotions" can cause a range of ailments. For example, a website associated with a healing and spiritual awareness center claims that "cancer is caused by the suppression of toxic emotions; primarily anger, hate, resentment and grief." Another website has a post from a certified hypnotist and psychospiritual therapist (you can get certified in that?) that suggests that pent-up anger ("emotional blocks") imprints on our cellular memory, causing a host of health problems.

Given all this pseudoscientific gibberish, it is no surprise that many people believe that pent-up emotions can result in cancer. A 2018 study that explored people's beliefs about the causes of cancer found that 43 percent of those surveyed believed that stress causes cancer. This was, as the authors note, the most commonly endorsed "mythical cancer cause." A 2015 study from Korea came to a similar conclusion, finding that the "most important perceived cause of cancer risk was stress."

Chronic stress—the kind associated with poverty, being subjected to ongoing conflict, lack of security, etc.—is linked to a range of health issues, including a weakened immune system. Stress matters, no doubt. But, as noted by the National Cancer Institute, "the evidence that it can cause cancer is weak." A 2013 analysis of all the available research, which included data from over 100,000 people, found no link between stress and colorectal, lung, breast, or prostate cancers. The few studies that have found a correlation are generally small and rely on the recall of the participants regarding their level of stress—which is not a reliable form of research.

But more relevant to a decision about whether to rant or not to rant: there is no science to support the idea that releasing your bottled-up rage in a cathartic burst is a helpful strategy for emotional or physical well-being. The idea of purging anger may have intuitive appeal, but sending a raging email, tweet, Facebook post, or graphic Instagram picture isn't a good way to deal with stress. On the contrary, it generally makes things worse.

We have known this for decades. Take, for example, the 2002 study with the title "Does Venting Anger Feed or Extinguish the Flame?" This study, which has since been replicated many times, found that using various strategies to vent anger, including punching a bag and blowing a loud horn at the person who was the source of the anger, did not help. In fact, the author, social psychologist Brad Bushman, concluded that "doing nothing at all was more effective than venting anger" and that the results "directly contradict catharsis theory." In short, ranting does not quell the anger, relieve stress, or make the ranter feel more emotionally stable. In fact, it has the *opposite* effect. It makes people feel *more* aggressive. In 2011, Professor Bushman said: "Venting anger is like using gasoline to put out a fire. It only feeds the flame by keeping aggressive thoughts active in memory and by keeping angry feelings alive."

More recently, researchers have explored the phenomenon of online ranting. A 2013 study, for example, found that reading and writing online rants—particularly on sites dedicated to the practice, known as rant sites—was "associated with negative shifts in mood." The researchers found that, in the short term, ranters felt better. So, although there is an immediate psychological reward for the action, in the long term, people felt worse. Professor Ryan Martin, the lead author of this study, told me, "Most of the evidence we have suggests that ranting just leads to greater problems down the road. Anger can be a healthy emotion, especially when expressed in effective ways, but ranting is rarely the most effective way."

Ranting is the junk food of emotional responses. A tasty McRant provides us with immediate satisfaction, but, with the passage of time, we often regret our behavior and might even feel a bit sick about it. One study found that 57 percent of Americans have posted something that they later regret. And, no surprise, many of these regrettable posts were made in the heat of the moment. Another study found that online users often experience a "social media hangover" that results in their deleting an offending post.

And like junk food, ranting can be seriously bad for our health. One series of studies, led by Elizabeth Mostofsky from the Harvard School of Public Health, explored the connection between expressions of emotion and cardiovascular health. One of these studies, published in 2013, found that the risk of experiencing an acute myocardial infarction (aka a heart attack) was "more than twofold greater after outbursts of anger compared with at other times, and greater intensities of anger were associated with greater relative risks." Think about that the next time someone tells you to let it all out.

From the perspective of emotional and physical well-being, ranting does not work and is likely harmful. But we shouldn't forget that ranting, particularly online ranting, can have other ramifications.

I am fairly active on social media. I don't have a celebrity-level number of followers, but I have enough that I am the target of many rants. I've been called an obnoxious ignoramus, a bigoted ignoramus, a toxic creature, a fear pusher, a mouthpiece for mainstream medicine, a soulless bastard, and a fucking idiot (the last one too often to count). Someone has even used the "C word" to describe me on Twitter. And I have often been told that I am a Big Pharma shill—because I critique unproven alternative therapies—and that I have a very small penis. The latter speculation was in response to an article I wrote denouncing the science-free bunk that is homeopathy. (But since homeopaths think smaller is better and more powerful, perhaps this is a homeopathic compliment? Thanks?)

The point is: I keep it all. I make collages of the insults and share them online and use them in public lectures as a way to explore what is driving the anger on particular issues. (Attribution usually removed from the posts in question.) Screenshots are easy. After a while, ranters often delete their nasty posts about me. But I almost always capture them before they are removed.

We should all assume that once a rant is posted, emailed, or made public in any tangible way, it will live forever. Because it will—even if you delete it.

That said, please keep the insults coming. They are gold.

Given that ranting and, more broadly, the entire idea of catharsis have never been supported by science, they have had a remarkable cultural run. While we don't hear much about primal scream therapy anymore, belief in the basic concept appears undiminished. What's going on?

There are many reasons for the enduring appeal—including the pop culture embrace of the theory and the mere fact that, short term, it feels good. But I think the biggest reason is simply a broad collective hunch that the concept of catharsis is correct. It is an elegant idea that accords with our view of how the world works. And it lends itself to the use of persuasive metaphors—the unhealthy festering of bottled-up rage—that reinforce the idea's intuitive attraction.

"It feels like a very nice and neat theory," Dr. Martin suggested when I asked him why the myth won't die. "The notion that we are these human pressure cookers that walk around building up steam until we blow—unless we let a little out—sounds like it would make some sense. In the end, though, it just isn't true."

This is a good reminder that just because something feels right—and feels good—doesn't mean it works. So when you are feeling that sense of rage and feel compelled to respond, what should you do?

First, you should let some time pass before you make a decision about responding. Try to calm down. Time will likely cool the rage and you can make a more dispassionate assessment. Don't tweet angry!

Second, and more important, think of a constructive way to harness your anger. "For whatever reason, the public seems convinced that our only options are venting or bottling it up," Dr. Martin told me. "But that isn't true. As human beings, we can express our anger in infinite ways—letters to the editor, polite assertion, exercise, art, journaling, writing music, poetry, protesting, and so on." Then again, there are times when bottling your anger might be the best course. Ranting to your boss, for example, is probably not the wisest way to address workplace anger. "We should think of our emotions as alerting us," Dr. Martin continued. "Fear tells us we're in danger, sadness tells us we've lost something, and anger tells us we've been wronged. What we choose to do when wronged is up to us and there are lots of healthy things we can do."

For now, back away from the keyboard.

1:30 PM—Thank-you letter

Instead of ranting, how about finding an excuse to write a thank-you note to someone? A study from 2018 found that people underestimate how positively the recipient will react. We also overestimate how awkward it would make the recipient feel (it almost never did). Indeed, research has pretty consistently found that saying thank you is good for both the sender and the receiver. It helps to build relationships and, no surprise, improves how the world sees you. (Not that that should be your motivation.) As the researchers of the 2018 study conclude, "Positive social connections are a powerful source of well-being, and creating those connections can sometimes come at little or no cost."

So dash off a brief note. Don't stress about the wording or penmanship. Just be sincere. For a small investment of time you can make your day and someone else's a wee bit better.

1:45 PM—Stand up

Admission: I have a standing desk. And I use it. And I love it.

I'm a bit embarrassed about this because being an early adopter of a hot health trend is antithetical to my very being. Whenever I hear about a new exercise, diet, or wellness strategy (it's increasingly about "wellness," isn't it?), I usually think, "Where is the evidence? Don't people know these things never pan out long term?"

But with the standing desk I jumped on the perpendicular posture train as soon as it left the station. Like so many of us, I spend a huge hunk of my day sitting on my ass staring at my computer screen. A device that injects some ergonomic variety sounded sensible. Being a science geek, I did do some prepurchase research—which was admittedly pretty superficial and likely influenced by a bit of confirmation bias, as I was already sold on the idea. Still, the research I found seemed supportive. Sign. Me. Up.

Despite this genuine enthusiasm for my fancy desk, when a colleague walks past my door and gives me that universal coworker-head-nod hello and sees me standing there like a tourist on a Segway, I feel a slight shudder of shame. But, damn, I love my desk. And who can blame me? Over the past few years standing desks have received a massive amount of positive press. And the danger of sitting—which has been called the new smoking—has been the subject of countless headlines.

The standing desk is far from new. Thomas Jefferson had a beautiful six-legged "tall desk" that can still be seen at his home, Monticello. Ernest Hemingway, Virginia Woolf, and Charles Dickens all wrote at a standing desk. And I am pretty sure Scrooge made that poor Bob Cratchit use one. But the standing desk as a broadly accepted wellness device is a recent development. In just a few years it has blossomed into a market that is expected to be worth almost $3 billion by 2025. A 2017 study from the

Society for Human Resource Management found that standing desks are the fastest-growing employee benefit in U.S. workplaces.

Part of their popularity is, of course, closely tied to the genuine health problems associated with being sedentary. According to the American Heart Association, sedentary jobs have increased 83 percent since 1950. A 2019 study concluded that since 2001 the prevalence of sitting to watch TV or use a computer at home has increased substantially. A growing body of evidence suggests this is not a good thing. Far too few of us get enough exercise. In 2018 the World Health Organization declared that physical inactivity was a global health problem, estimating that "3.2 million deaths each year are attributable to insufficient physical activity." The marketing for standing desks leverages these concerns and bundles them with other alleged benefits, including weight loss and improved concentration.

But do they work? Do they improve your health or promote less sedentary behavior?

Professor Stuart Phillips is a renowned physiologist at McMaster University's Department of Kinesiology. "The most recent evidence," he told me, "shows that the presence of a standing desk in the workplace actually does a little to reduce people's sitting time, but not by much." The research he was referring to includes a large 2018 systematic review of the evidence that concluded, "At present there is low-quality evidence that the use of sit-stand desks reduce workplace sitting at short-term and medium-term follow-ups." More importantly, the study notes, "there is no evidence on their effects on sitting over longer follow-up periods."

For Professor Phillips, this last point is key. "What are the effects of this on long-term health? We just don't know," he told me. "To this point we really can't say too much about what standing desks do." He is not saying that sitting a ton is a good thing. There is lots of evidence—mostly, but not entirely, from correlation studies that fail to establish causation—that sitting is problematic. A 2018 study from Maastricht University, for example, found that excessive sitting isn't great for the metabolism. (The researchers had participants sit for 14 hours a day, with bathroom breaks as the only activity.) Both exercise and standing benefit our cardio and metabolic health, but in different ways. This suggests, to quote the authors of the study, "the need of both performing structured exercise as well as reducing sitting time on a daily basis."

Professor Phillips agrees. "Being sedentary and not physically active is a bad combination." But while there is universal agreement—and a boatload of good research—to support the profound benefits of physical activity, Phillips thinks the role of sedentary behavior is more complicated. There clearly seems to be a connection between sitting and bad health. A 2018 study of over 120,000 people found prolonged sitting to be associated with a range of health issues, including cancer and cardiovascular disease. But teasing out the benefit of standing at work—how much, how often—to our long-term health isn't easy. For example, the above-noted 2018 research review concluded that standing desks can lead to about an hour of less sitting at work, which sounds pretty good. But it also concluded that it remains unclear whether this would have any health benefit.

Still, some of the science suggests standing desks have promise. A large 2018 study from the U.K. that provided people with an adjustable desk, ongoing feedback, prompting to stand, and one-on-one coaching did result in objectively measured reduced occupational sitting time and improved job performance (though no increase in overall physical activity). Yet it's unclear whether this kind of intensive intervention could be applied outside the research setting. How many employers are going to provide standing prompts and coaching?

We should also recognize that standing *too much* can be harmful. A 2017 study examined data from over 7,000 employees and found that "occupations involving predominantly standing were associated with an approximately 2-fold risk of heart disease compared with occupations involving predominantly sitting." You may think this is obvious because standing jobs are associated with lower socioeconomic circumstances and, as a result, poorer health. But even when controlling for these variables, the researchers still found that standing too much was associated with increased health risks. There is also some evidence that, for some people, standing too much can result in potentially harmful muscle and joint strain.

Now, I'm not convinced that studies like this necessarily negate the potential benefits of a standing desk, but they do highlight how confusing and equivocal the evidence is. It could be that both too much and too little standing are bad for your health. To give you a sense of how unclear things currently sit, standing-desk-wise, consider the advice provided by Professor Keith Diaz, a behavioral medicine researcher at Columbia University Medical Center, on the value of standing desks. "For those individuals who are looking to reduce their sitting time at work, the best scientific evidence we have at the moment suggests that sit-to-stand desks may be effective and could reduce your sitting time at work." But then he added a rather large qualification. "It is worth noting that we still don't know if standing is a healthier alternative to sitting."

There is yet another wrinkle: might standing desks cause people to exercise less? Exercise works. Exercise is essential. We all know this. But perhaps some people will feel that since they used a standing desk at work, a practice that studies have shown does not require that much physical exertion (and, by the way, does not burn many calories), they won't feel as compelled to exercise at other times in the day. This is called compensation behavior.

Consider a 2016 study from the U.K. that followed office workers for three months while they used standing desks. The desks did cause the participants to reduce their sedentary behavior at work. But here is the kicker: "These changes were compensated for by reducing activity and increasing sitting outside of working hours." As a result, there was in fact no increase in nonsedentary behavior, and the findings hint that standing desks might invite some people to actually exercise less. Other studies, though, did not find compensation behavior, so this remains a big and important unknown about the health value of such desks.

In many ways, this kind of result shouldn't surprise us. Meaningful long-term behavior change is extremely difficult, particularly since humans appear to be hardwired to prefer sedentary behavior. For most of human history, there was an evolutionary advantage to avoiding the expenditure of energy. When calories are hard to find, you don't want to burn them unless it's absolutely necessary. A 2018 study from the University of British Columbia examined brain activity to get a sense of the degree to which we are, literally, hardwired to prefer sitting to standing. The research involved monitoring the participants' brain activity while they considered various scenarios involving both sedentary and active behaviors. What the researchers found was that, yep, more brain activity is required to overcome the tendency toward sitting. Specifically, the researchers concluded that "sedentary behaviors are innately attractive and that individuals intending to be active need to activate cortical resources to counteract this innate attraction." Simply put, at an unconscious level, our brains are constantly telling us to take a load off.

Standing desks are a good example of how an intriguing but pretty convoluted body of evidence can be twisted by various forces, including marketing pressure and media hype, to create an impression that a health intervention has more value than it actually does. And the media hype has been significant. A 2017 study from the University of Sydney, entitled "Overselling of the Sit-Stand Desk," concluded that media reports overplayed both the potential harms of sitting and the health benefits of the standing desk. The study also found that more than one-third of the relevant newspaper articles suggested that sitting cancels out the benefits of exercise. Some emerging research does suggest that prolonged sedentary behavior may cause physiological changes that blunt the benefits of exercise. But physical activity is always a good idea. Sitting a bunch won't erase that reality.

And by the way, sitting is *not* the new smoking. I believe that this provocative claim—which will get you a huge number of hits on the internet—is one reason this topic has received so much attention. It is catchy and scary. It makes for a terrific headline. But it is also wrong. A 2018 study published in the *American Journal of Public Health* analyzed all the available evidence about the health risks of both smoking and sitting—with the specific goal of exploring the veracity of the "sitting is the new smoking" slogan. What the researchers found is that there is simply no comparison. Smoking is vastly more harmful than sitting. As the lead author of the study, Professor Jeff Vallance, said, "The risk estimates for smoke are clearly in a different ballpark."

To put the sitting vs. smoking risk in perspective, consider these numbers. An observational study found that there is a 44 percent increase in colon cancer risk for those who sat the most at work compared with those who sat the least. That sounds scary. But this is just one study (that is, we need more research to confirm the finding) and, more importantly, it is nowhere near the risk increase associated with smoking. According to the U.S.'s National Cancer Institute, the lifetime risk of getting diagnosed with colon cancer is 4.2 percent. So those who do the most sitting in the workplace increase their lifetime risk from 4.2 percent to about 6.2 percent. If this research is consistently replicated by other researchers, that is a meaningful increase, but not close to the impact of smoking. According to the Centers for Disease Control and Prevention, "People who smoke cigarettes are 15 to 30 times more likely to get lung cancer or die from lung cancer than people who do not smoke." The World Health Organization puts the risk even more bluntly: "Tobacco kills up to half of its users."

The hyperbolic descriptions of the risk associated with sitting do little to inform a dispassionate public discourse on how best to address what is a real public health concern. It does, however, help to sell standing desks.

Like so many others, I fell prey to the persuasive recipe of media hype, intuitive appeal, fearmongering, marketing pressure, and a dash of real, albeit still evolving, science. We have seen variations on this formula throughout this book. But the story of the standing desk is also a good example of how easy it is to be deceived by a selective review of the evidence. A quick scan of the science associated with standing desks can produce studies and media reports that appear to provide pretty unwavering support. Those were the articles I found when I first jumped on the standing desk bandwagon, likely because I was keen to find exactly those kinds of articles. But, as is so often the case, the story is actually much more complicated.

Fixing a complex public health problem like sedentary behavior is rarely accomplished by a single intervention, especially when that intervention requires people to put effort into taking action. One analysis from 2016 found that health behaviors that require people to use a high level of agency—that is, personal effort—tend to be favored by governments (mostly because they are cheaper and less politically challenging), but they don't work as well as "interventions that require individuals to use a low level of agency to benefit." In other words, public health interventions, like simply asking people to be more active, do not work as well as structural changes, like building cities that encourage an active lifestyle. Alas, our society tends to gravitate to health fixes that place the responsibility on the individual. It is easier to sell products—like standing desks—than to alter how our streets are constructed.

"We need a broader social and environmental approach to address the decreasing levels of physical activity in our daily lives," Professor Josephine Chau told me. She is senior lecturer in public health at Macquarie University in Australia and the lead author of the study on how the media covered standing desks. "When we think about all the sitting contexts in everyday life, like the workplace or school, at home, during travel, and in leisure time, it's logical that a single approach would not be enough to cover all of these."

Standing desks are unlikely to be a magical solution for anything. The research suggests they probably help a bit, but a more global approach is needed for both individuals and society. The decision you need to make is simple: be more active, however you do it. "Get up and move!" Professor Stuart Phillips said. "Take a break, go for a brisk walk, stretch, and try and reduce sitting time. But, and this is a big one, be physically active outside of your sedentary office job. It's the person who isn't physically active and sedentary at work who has the greatest risk."

For now, I'll keep my desk. Despite the lack of conclusive evidence, all the science-twisting media hype, and the office-dork vibe, I like the variation the desk gives me. A bit less sitting certainly isn't a bad thing. And if it was good enough for Thomas Jefferson and Bob Cratchit, it's good enough for me.

2:00 PM—Another coffee?

That post-lunch lull often hits hard. You are feeling sluggish. You've got a bit of brain fog. Your eyelids feel like lead weights. Is it time for another coffee?

For me, the answer is almost always yes, because, well, coffee. But researchers at the U.S. Department of Defense have tried to determine the exact timing required to optimize the benefit of caffeine intake. (Weaponized coffee consumption? I get that. God bless the need for alert soldiers. I picture geared-out Army Rangers sipping espressos in a government laboratory.) Their study, published in 2018, found that to maximize the anti-sleepiness potential of coffee, consumption should be timed to fit our chronotype—that is, those individual sleep patterns discussed at the start of this hypothetical day. By analyzing and testing various dosing strategies and a host of biomarkers, the researchers developed a coffee "optimization algorithm" that maximized the utility of "caffeine countermeasures" in the ongoing war against afternoon sleepy time. The researchers concluded that by targeting your caffeine intake you can increase your alertness, as measured by cognitive performance, by as much as 64 percent.

While that is a seriously impressive number, it isn't entirely clear from the study how the algorithm would play out for the non-combat-zone average citizen. The message seems to be to use coffee when it is needed most—and this is often from midafternoon to the end of the day. The research also suggests you could reduce caffeine consumption during times of natural alertness—for many, this might be midmorning. Aim to increase the intake right before the times you usually feel sluggish.

My guess is that the Department of Defense would approve a decision to consume some coffee at this time. Drink up, soldier.

Volume is another factor that may influence your decision to have a midafternoon caffeinated beverage. How many coffees are too many coffees? If you believe a 2018 study published by a team from the National Cancer Institute, a person can tolerate a lot of coffee—perhaps as many as eight cups a day.

Let's pause to consider where this conclusion comes from. This study was done using data from the huge UK Biobank cohort study, a research platform that involves demographic, lifestyle, genetic, and health information from half a million people. This massive repository of health data allows researchers to follow individuals for years to explore how a range of variables, such as drinking coffee, might relate to health outcomes. In the U.S. coffee study, the researchers looked at data for a period of 10 years and concluded, as have many other studies on point, that "coffee drinking was inversely associated with mortality, including among those drinking 8 or more cups per day." And this "coffee is good for you" conclusion seemed to hold regardless of a person's particular genetic predispositions.

It was a large, impressive, and well-done study. And, naturally, it led to many headlines. The media loves studies that say coffee (or chocolate or wine or beer or pasta or pizza or sex) is good for you. There were headlines like "Drinking 6 to 7 Cups of Coffee a Day Might Not Only Keep You Awake, It May Also Make You Live for Longer" and "Coffee May Help You Live Longer, Say Actual Scientists."

But despite the size of this study, it does not establish causation. Indeed, the researchers were careful to highlight this fact in the paper, noting that their work should be "interpreted cautiously" and that it only provides "further evidence that coffee drinking can be part of a healthy diet and may provide reassurance to those who drink coffee and enjoy it." To get a sense of causation—that is, that drinking coffee can cause a drop in mortality—the researchers would need to run a clinical trial that, at a minimum, randomly assigns one group to consume coffee and one to not drink coffee, control for a host of variables, and then follows the research participants for years, if not decades. This kind of study simply isn't practical.

Gary Schwitzer is an adjunct professor at the University of Minnesota School of Public Health and the publisher of the award-winning HealthNewsReview. This publication, which is one of my absolute favorites, does careful, science-informed critiques of health and science reporting. In his 45-year career in the area of health journalism, Gary has seen a lot of bad health reporting. I asked him how often does the media get causation wrong.

"Our experience shows that this is *the* most common flaw in health-care news," Schwitzer told me. "I don't think a day goes by without some news stories—reported by many news organizations—inaccurately using causal language to describe observational data."

Studies on media reporting confirm his impression. To make matters worse, the media seems to prefer the less methodologically robust observational research. A 2014 study that examined what kind of health research top newspapers cover discovered that newspapers were more likely to write about observational studies rather than more informative randomized controlled trials. It also found that "when the media does cover observational studies, they select articles of inferior quality." To compound the problem, the media rarely discusses the inherent limitations of observational studies. A 2015 study found that only 19 percent of stories about observational studies discussed the limitations of the methodology.

Schwitzer believes that this poor handling of observational studies "leads people to lose confidence in science, especially when it falls into the on-again, off-again back-and-forth tennis game of 'coffee is a killer' one day followed by 'coffee boosts protection against x' the next day. The public then becomes numb to all health care news, understandably rolling their eyes and feeling like it's all a waste of time. And much of it is."

Evidence supports Schwitzer's concerns. A 2014 study found that exposure to conflicting media reports on the health benefits of food, including coffee, was associated with confusion about what foods are best to eat. (This study critiquing conflicting nutrition research is also an association study! As with much of the nutrition research, it found a correlation. So we need to take care not to overinterpret the data. Again, this doesn't mean it is irrelevant, but rather that it is more suggestive than definitive.) More worrisome, such exposure can create the impression that scientists arbitrarily change their minds, leading people "to doubt nutrition and health recommendations more generally."

Are observational studies useless? Not necessarily. The key, Schwitzer believes, is recognizing their limitations. "Such research can have value," he told me. "Our knowledge of the harms of smoking came from observational studies. Our knowledge of the early risks from Vioxx came from observational data." And in some areas, observational research might be all that we will ever have. Still, it is best to think of these observational studies as rarely, if ever, definitive. They are almost always just one piece of an accumulating body of knowledge. When reading an article about a new study claiming something has a health benefit or harm, pay attention to whether it is an observational study. (Look for phrases like "associated with" or "linked to" or "correlated with.") If it is, assume that nothing has been proved and more science is likely required.

What does this mean for your decision about how many coffees to consume? Schwitzer noted that although "coffee has become the poster child of miscommunicated observational research," he believes it is safe to consume, even in relatively large amounts. "The weight of evidence (albeit observational) is piling up on the side of coffee's potential benefits," he said.

So, once again, it is probably a fine, and perhaps even a somewhat healthy, decision to grab another coffee.

2:15 PM—Soap

On average, people need to pee four to seven times a day. And since you have already consumed several cups of coffee, you will likely need to pee several times this afternoon. You are going to wash your hands afterward, of course. But do you decide to use soap or hand sanitizer?

Studies have consistently confirmed the value of using soap to wash your hands, especially if you wash them properly. There is no need for antibacterial soap. Given the growing concern about antibiotic resistance, antibacterial products are—when you are thinking long term and on a population scale—a bad idea. Indeed, after studying the issues, the FDA recommended against the use of antibacterial soap and called for a ban on the marketing of products containing particular chemicals. They did this for several reasons, including concern about antibiotic resistance and the fact that there is no evidence to support the use of such products. Dr. Theresa Michele, director of the FDA's Division of Nonprescription Drug Products, put it bluntly in an agency press release: "If you use these products because you think they protect you more than soap and water, that's not correct."

Soap is a pretty miraculous product. The Babylonians are said to have been the first to perfect its production, around 2800 BC. Since then it has become central to the hygiene habits of almost every civilization on the planet. If used appropriately, it is a simple and tremendously effective public health tool. Indeed, there is now a Global Handwashing Day (October 15), promoted by bodies throughout the world, including the World Health Organization. And organizations like the Global Soap Project recycle soap from hotels and businesses and send it to communities that are in need.

Derreck Kayongo is the founder of Global Soap Project. "An estimated 1.8 million children under the age of five die each year from diarrheal diseases and pneumonia, the top two causes of death among young children worldwide," he told me. "Handwashing with soap can reduce the incidence of diarrhea among this age group by approximately 30 percent and respiratory infections by approximately 20 percent. Soap is a prerequisite to informed public health."

Kayongo lived through a civil war in Uganda, and his experiences in a refugee camp in Kenya inspired his efforts, which won him a prestigious 2011 CNN Heroes award. When he first got to the U.S. he stayed in a hotel, where he was amazed by all the soap. His room had hand soap, shower soap, and shampoo. And much of it was simply thrown away. Kayongo recognized an opportunity.

In addition to soap's ability to save lives, Kayongo noted that it has a broader social impact. "Think about it . . . soap costs about just fifty cents, but treating diarrheal diseases can cost as much as $7, which is way above the per capita incomes in the global south. So soap is clearly tethered to the economic performance of each community."

Given his passion for all things soap, is Kayongo a good hand washer? Does he walk the soapy talk? "Ever since I built the Global Soap Project I have become a fanatic when it comes to handwashing. So yes, I am an avid hand washer."

I asked him if he ever rushes or skips the process. His response: an emphatic "NO!"

Think about Kayongo and his project the next time you feel like saving yourself 20 seconds of time in the bathroom.

Most public health agencies recommend washing your hands with soap and water several times a day, including after using the bathroom, before and after preparing meals, and after touching garbage. Although alcohol-based hand sanitizers are not as effective as soap and water, in a pinch they can be a useful substitute. Studies have found that hand sanitizers (which should contain at least 60 percent alcohol) are effective at killing many, but not all, pathogenic germs. However, they can't remove other potentially harmful substances, such as a dangerous chemical, from your hands. If your hands are grimy, good old soap and water is the best approach. Unlike hand sanitizer, soap and water works by washing away germs, not killing them. So, if done well, it is a more comprehensive approach.

But what if the soap is dirty? *Can* soap have germs? There is a famous scene in the TV show *Friends* where Joey and Chandler are having a roommate fight about this very question. Joey: "Why can't we use the same toothbrush but we can use the same soap?" Chandler: "Because soap is soap. It's self-cleaning." Joey: "All right, well the next time you take a shower, think about the last thing I wash and the first thing you wash."

Is Joey making a good point? Should we be wary of well-used bars of soap in a public washroom? There doesn't appear to be a lot of independent research on the matter, but Chandler's view is close to the truth (though Joey gets points for imagery). In a 1988 study, researchers infected commercial soap with pathogenic germs. They found that the soap did not transfer the bacteria and concluded that "little hazard exists in routine handwashing with previously used soap bars." At the risk of confusing your decision-making process here—which is antithetical to the goal of this book—a few studies have found that public restroom liquid soap dispensers can become contaminated with germs if they're not cleaned properly.

In the end, we should all strive to be, like Derreck Kayongo, avid hand washers. We can never reduce risks to zero. But looking at the evidence of the benefits, the odds clearly favor the frequent use of our 5,000-year-old friend, soap.

2:17 PM—Drink water

Sure. If you are thirsty, have a drink of water. If you aren't, don't.

The idea that we are all supposed to drink some prescribed amount of water is one of the most persistent health myths. You do not need eight glasses of water a day, unless you feel you need eight glasses a day (that is to say, you are thirsty). After eons of evolution, the human body is darn good at regulating fluid requirements. You don't need a smartphone app to remind you to drink. And unless you are about to trek across the Gobi Desert, you probably don't need to carry around a $50 stainless steel water container. When the need for fluids arrives, you can just drink good ol' tap water. (At work I drink water out of the faucet in the office's kitchen sink, kitty style.) Yes, there are jurisdictions where water quality has been lacking or dangerous—Flint, Michigan, is a notorious recent example. But for the most part you can ignore the entire multibillion-dollar H_2O industry. Water is just water. Drink as needed.

Here are some of the other water-related things you don't need.

Expensive bottled water. There is a luxury brand of water called Acqua di Cristallo Tributo a Modigliani that is worth $60,000. It comes in a handcrafted 24-karat gold bottle that contains a combination of glacier water from Iceland and natural spring waters from France and the Fiji Islands. It will not hydrate you any better than the water that flows out of the tap in a McDonald's bathroom. (But what are you doing at McDonald's?) In fact, unless there is good reason to be worried about the tap water where you live or you simply don't have access to clean water, there is no health-related reason to buy bottled water. In addition, a few studies have found that bottled water—a lightly regulated industry—contains more bacteria than tap water. And blind taste tests have revealed that people usually can't tell the difference between bottled and tap, and some actually prefer the taste of tap water over premium bottled water.

Alkaline water. A bevy of wellness products claim to reduce the acidity of your body because being too acid is bad, or so the marketing goes. The human body has evolved to keep your pH within a healthy zone. Your body constantly and tightly regulates the pH of your blood to maintain a state that is already fairly alkaline. You don't need a trendy new wellness product to help that natural biological process. As McGill chemistry professor Joe Schwarcz has noted, you can't change the pH of your body with the food you eat and the fluids you drink—and if you could, it would kill you. In addition, there is no evidence to support the health claims made by the alkaline water industry. A 2016 review of the evidence, for example, found that "despite the promotion of the alkaline diet and alkaline water by the media and salespeople, there is almost no actual research to either support or disprove these ideas."

Vitamin water. Get your vitamins from the food you eat. You don't need gimmicks like vitamin water to supplement your diet. Also, drinking Vitaminwater, the Coca-Cola product, should not be seen as the same as consuming water. It is a sugar drink filled with empty calories. Avoid.

Oxygenated water. This one is ridiculous on many levels. How, exactly, do the manufacturers of hyper-oxygenated water, as it is often called, inject the extra oxygen into their product? If they are changing the actual molecular structure of the water, then it isn't water anymore. If you add an oxygen molecule to water, you get H_2O_2, also known as hydrogen peroxide—which, if consumed, is toxic. A small amount of dissolved oxygen is present in water (that's how fish breathe), and if you put it under extra pressure, water will allow some additional oxygen molecules to be dissolved. But release the pressure (as in open the bottle) and it leaves, just like carbonated water. Most important, how is your body supposed to absorb the extra oxygen? Your gut doesn't operate like your lungs. Your intestine doesn't breathe. And you aren't a fish.

Gluten-free water. Not a thing.

Organic water. Not a thing.

GMO-free water. Not a thing.

Raw water. This is a thing, and it is completely nuts! Over the past few years, several companies have been marketing the idea that untreated water, often straight out of a stream or the ground, is especially healthy because it is more "natural" (there's that reliance on the conceptually incoherent naturalistic fallacy) than water that has been disinfected and purified. Live Water, one of the best-known companies in this space, claims that its product contains healthy bacteria that are lost in the purification process. This product—basically just dirty water in a fancy glass jug—sells for around $60. The reality, of course, is that clean water is one of the single greatest public health achievements in the history of humanity. In 2010 the United Nations declared access to clean water a universal human right. Yet over 800 million people still lack access to clean drinking water, a problem associated with a host of diseases, including diarrhea, cholera, dysentery, typhoid, and polio. The World Health Organization estimates that contaminated drinking water causes more than half a million diarrheal deaths each year. In this light, the raw water trend is especially worrisome, if not perverse, as it so clearly runs counter to the science that has saved millions of lives.

All this water gobbledygook is a good example of how something so basic to our health—drinking water—can be twisted by market forces and bad science. But I do like the emphasis on drinking water over things like sugar-loaded pop, sports drinks, and fruit juices. If you are thirsty, water is the best thirst-quenching choice.

2:20 PM—Office meeting

It has been estimated that meetings—the ultimate soul-crushing workplace time waster—cost the U.S. economy $37 billion each year. There are 11 million meetings in the U.S. every day, and 37 percent of them start late, mostly due to tardy attendees (probably me). The average employee spends between five and seven hours a week in meetings (and for CEOs meetings account for 72 percent of their total work time), many of which are either totally useless or just superinefficient. Even when you think you are being productive, you probably aren't. One study found that premeeting banter makes people later *think* the meeting went great, when a more objective analysis demonstrates that, nope, the meeting was just as unproductive as meetings without the office gossip. And when asked, managers from a variety of industries put the productivity of meetings at a dismal 33 to 47 percent.

Research has also found that over the past few decades the frequency and length of meetings have increased. For most of us, meetings are like the Borg from *Star Trek*: a mindless, relentless force competing with other tasks (mostly email, as it turns out) to consume our entire working life. Your time is almost always better spent doing other things. Unless you absolutely must go to a meeting, do whatever you can to get out of it and focus on some actual work.

In fact, just thinking about a meeting can be a drain. As noted earlier, an upcoming scheduled event—like a meeting—creates the impression that you have less time than you actually have. When an event looms, you feel busier and do less productive work. So meetings make you less productive before you go and waste time while you are at them. Talk about a lose/lose situation.

Teleconferences are no better. Absolutely no one on the call is actually listening. One survey, from 2014, found that 65 percent admit to doing other work while on a teleconference. Another 55 percent say they also eat or make food; 47 percent admit to going to the bathroom (while on mute, I assume); and 6 percent make other calls. (Remember that, overly-keen-conference-call-chairperson who takes attendance. No one on the call is taking this seriously.) Teleconferences, then, are even worse than face-to-face meetings from the perspective of being meaningfully productive. They have been reduced to a strange we-gotta-do-this ritual that usually accomplishes almost nothing.

While we're at it, let's debunk one of the biggest justifications for meetings: brainstorming. Getting together to bounce around ideas has long been a core reason for the existence of long, rambling meetings. An advertising executive named Alex Osborne made the concept popular in the 1950s. Osborne is given credit for the term "brainstorming," which he believed increased both the quality and quantity of good ideas by allowing a free, uninhibited flow of creativity. Many minds working collaboratively together are better than one working alone, was the premise. But while it sounds logical, it does not, in fact, work. As noted in a 2003 review of the evidence by psychology professor Adrian Furnham of University College London, "Research shows unequivocally that brainstorming groups produce fewer and poorer quality ideas than the same number of individuals working alone." Another review, done in 2002, came to the same conclusion: "Much literature on group brainstorming has found it to be less effective than individual brainstorming." What this means is that it is more effective to think up ideas on your own and then simply share them. No meeting required.

It is amazing, then, that we still hear so much about brainstorming. The reason has popped up again and again in this book. It is a practice that fits with how we *think* the world works, even if the world doesn't work that way. As noted by psychology professor Tomas Chamorro-Premuzic in a 2015 article in the *Harvard Business Review*, "Brainstorming continues to be used because it feels intuitively right to do so." But the world isn't flat, and brainstorming doesn't work.

Teamwork and cooperation are valuable and can certainly be productive. And interesting research is emerging on how to encourage effective collaboration, including building work environments that promote communications (often a chance meeting in a hallway can be more fruitful than a scheduled meeting). But don't fall for the brainstorming bunk.

Interesting fact: the open workspace office—an approach that was justified, in part, by the idea that it would increase constructive interactions among employees—also does *not* work. In fact it seems to have the exact opposite effect. A 2018 study found that face-to-face interaction decreased 70 percent in the open workspace environment. The authors summarize their findings thus: "Rather than prompting increasingly vibrant face-to-face collaboration, open architecture appeared to trigger a natural human response to socially withdraw from officemates and interact instead over email and [instant messaging]." It's tough to force people, either through scheduled meetings or wall-less work environments, to have creative interactions.

If you must have meetings, limit the number of them per week, make them as short as possible, end on time, and for the love of God, have a clear purpose. Also, consider different formats, like standing, walking, or dancing on the spot. Let me know how that goes, because I won't be there.

2:30 PM—Nap time

The beds looked seriously comfy. That surprised me. I had expected military-style fold-out cots or, at best, those squeaky air mattresses your friends' parents made you use at sleepovers. But the big room was filled with little beds, perhaps 20, that had freshly cleaned sheets and great-smelling pillows.

The instructor—a sleep "coach"—dimmed the lights and had us do some light stretching and a few breathing exercises, and then he asked us to commence the main activity, sleep. We all climbed into our cozy beds and closed our eyes. It was midafternoon, and I was in a room filled with new parents, stressed-out students, and busy office workers taking a nap. The last time I experienced a communal nap time was in kindergarten. But this wasn't some strange adult reenactment of foundational education experiences. It was a napercize class. This was a room full of people paying to sleep.

Napping has gone mainstream. Once frowned upon as a practice of the lazy and unmotivated, it has been embraced by the corporate world as something that should be encouraged. In a short amount of time we've gone from viewing napping as something that could get you fired, to something many business leaders, most notably Arianna Huffington, think more of us should be doing—perhaps every day. Many companies now have sleep rooms where employees can escape to get some shut-eye. Google, for example, has installed sleep pods—sci-fi cocoons that play relaxing music—in their offices that are specifically designed for naps.

My napercize experience is part of a broader sleep-industry trend capitalizing on the recognition that sleep is key to good health. A 2017 market report estimated that the sleep-health industry is worth between $30 billion and $40 billion a year, and that it continues to "grow by more than 8 percent per year, with few signs of slowing down." We have our choice of a host of nap-enhancing products, including the ostrich pillow, a padded device that covers your entire head and is designed to help you sleep anywhere, such as at your desk or in a work cubicle. There are adult-sized fleece onesies, inflatable napping pods, desks with hidden nap nooks, nap-inducing eye massager goggles, and office chairs that collapse to become beds.

There is absolutely no doubt that many of us need more sleep. Good sleep is an important part of a healthy lifestyle. Sleepiness on the job can result in accidents, poor concentration, and reduced productivity and creativity. Studies have shown that highly sleepy workers are 70 percent more likely to be involved in accidents than well-rested workers. Some of history's most dramatic and tragic industrial accidents, such as the Chernobyl and the *Challenger* space shuttle disasters, have been associated with a lack of sleep. And numerous studies have found, not surprisingly, that feeling sleepy can significantly lessen productivity. A 2010 study of over 4,000 employees found that those categorized as being sleep deprived had "significantly worse productivity, performance, and safety outcomes." In addition, the researchers found that "fatigue-related productivity losses" were estimated to cost the employer $1,967 per employee annually. So it is understandable that many companies may want their employees to take naps. It is good for the bottom line.

Given this rising cultural acceptance of napping and the adverse consequences of workplace sleepiness, you might suspect that the conclusion to this section would be a straightforward declaration that you should put your head down on your desk right now and catch some z's. Unfortunately, it isn't that simple.

"Okay, so naps are like cupcakes," said Professor Colleen Carney. "I love cupcakes, especially in moderation, but it's not like they are good for you, nor are they a disaster."

Professor Carney is the director of the Sleep and Depression (SAD) Laboratory at Ryerson University. She has spent her career studying sleep, so she is obviously thrilled that its value is now being taken seriously. However, she is not impressed with some of the simplifying hype that now permeates popular culture. "People love to talk about naps, and there is a growing interest in trying to do it 'successfully,' something I find obsessive."

Why this reaction? Despite all the current enthusiasm, the evidence about the benefits of napping is not nearly as conclusive as is often portrayed. Yes, naps can give you a boost. A short nap (more on shortness in a bit) can enhance learning ability, performance, and wakefulness, which can all be advantageous in work situations. A classic study by NASA found that a brief nap improved performance by 34 percent. Other studies suggest that napping may improve cognitive function, including among older individuals with cognitive impairments.

But napping, particularly for longer periods, has also been associated with some negative outcomes. A 2016 analysis of studies, which involved data from more than 100,000 people, found "a significant association between daytime napping and hypertension." Other studies, some quite large, have revealed an association between daytime napping and an increased risk of type 2 diabetes and heart disease. Much of the research that has found harm is correlational in nature. That is, we are not sure napping causes the problems. It might be that people who nap are simply more likely, for other reasons, to develop these health conditions. And other studies have found that the ratio of benefit versus harm could depend on a number of factors, including the length and quality of nighttime sleep.

All this research highlights the somewhat confused nature of the evidence. As summarized in a 2017 academic review aptly titled "Exploring the Nap Paradox: Are Mid-Day Sleep Bouts a Friend or Foe?" the totality of the research seems to suggest that napping of a particular duration can, in certain circumstances, have benefits for young, healthy individuals. But for some populations, including older adults, "excessive napping has been linked with negative outcomes."

Sports are one area where the emerging conventional wisdom doesn't fit what the science says. Although plenty of commentators suggest that athletes should use naps to enhance performance, the evidence is, once again, less than definitive. For example, a small 2014 study of young athletes found that "napping showed no reliable benefit on short-term performances of athletes." Dr. Andrew Watson, a sports medicine expert at the University of Wisconsin and the author of a 2017 review of the literature, concludes, "The role of daytime naps on performance is unclear."

This divide between the emphatic napping endorsements and the actual sports science doesn't surprise Dr. Charles Samuels. He is one of the sleep experts we met at the start of our hypothetical day. Sleep and sports make up his specific area of expertise, so he knows this literature inside out. Part of the problem, Dr. Samuels told me, is that, in the context of sports, it is tough to get the needed data. "Sport science does not receive the kind of funding necessary to do the epidemiological studies to give us the kind of evidence to make substantial claims," he noted. In addition, he said that the weakness of the science does not get enough airtime. "The quality of the evidence is rarely discussed properly, and in sport science the limitations sections"—the part of academic papers where study limitations are disclosed—"are notoriously weak."

Dr. Samuels has no doubt that when it comes to athletic performance, "sleep optimization" strategies help. Good sleep and the right amount of rest matter, even for non-elite athletes (that is, 99.99 percent of us).

But, he concluded, "there is not enough high-quality evidence to suggest that it is a performance enhancement technique."

The biggest reason many experts have a more muted response to napping than what's heard in popular culture is that it may disrupt your nighttime sleep patterns. Most people who nap do so because they feel sleepy, and that sleepiness may be due to a lack of quality sleep at night. Poor sleep is—no surprise here—associated with an increased likelihood of napping. Sleeplessness = sleepy.

Napping can create a vicious cycle for people with insomnia. According to Professor Carney, "In the context of insomnia, [napping] is a really bad idea because it takes away an exponential amount of deep sleep at night." A person who can't sleep at night may feel compelled to take a daytime nap (and let's not forget, pop culture keeps telling us that napping is a good idea). But this will make it more difficult to sleep at night, which, in turn, will make the sleepy person want to nap the next day. And round and round it goes. For these people—and it has been estimated that approximately 30 to 50 percent of us suffer from at least brief periods of insomnia—napping is something that should be avoided. As summarized by the U.S.'s National Sleep Foundation: "If you have trouble sleeping at night, a nap will only amplify problems."

In addition, napping can result in what is called sleep inertia—that groggy, confused sensation that we often feel immediately after waking. Although this usually doesn't last long, particularly if the nap was short, it can have implications for those who must perform complicated or dangerous tasks immediately after waking. You don't want a brain surgeon, airline pilot, or anyone carrying a firearm making key decisions in a post-doze daze.

So, where does all this this leave us? Professor Carney suggests thinking of naps as a tool to alleviate sleepiness for a particular purpose—you may have an important activity to perform, like a big talk, a tough assignment, or an exam—but not as a primary sleep strategy, especially if you have trouble sleeping at night. "Napping is a good idea when it is more important to manage daytime sleepiness than to protect nighttime sleep," Professor Carney said. "Otherwise naps are unimportant and not great for sleep, but fine for an indulgence—just like a cupcake."

Our current enthusiasm for all things nap related illustrates how cultural momentum can build around an idea—as we saw with the standing desk—resulting in the distortion and oversimplification of the underlying science. This can in turn feed a kind of it's-a-fact mythology that can misinform our decision-making.

Consider this nap-related example. In 2015 a study was published suggesting that taking a nap could reduce frustration at work. The research generated a bunch of media attention, with headlines like "Feeling Angry? Take a Nap." While the conclusion of the research makes intuitive sense, the research was done by a PhD student and was just a pilot study—a small, preliminary investigation designed to explore the value and feasibility of a larger, more methodologically robust study. Pilot studies should not be held up as a source of definitive evidence. I could find no indication that this pilot study has been replicated or that a larger version has been completed.

Despite this, the idea that napping reduces frustration is everywhere, and this study is often cited as the source of what is now taken as a truism. I'm not saying that the finding isn't, necessarily, correct. But this study is far from conclusive evidence. In fact, a pilot study is by definition *not* conclusive. And when this study is noted in the popular press—which is often—the article almost never mentions that it was a small pilot study.

A combination of genuine need (sleep *is* important!), market forces, cultural momentum (thanks, Arianna Huffington), and oversimplified science has elevated the status of the nap, creating the impression that it's a key part of a healthy lifestyle.

One of my goals for this book is to simplify your daily decision-making, but, unfortunately, napping is one area that is, as Carney notes, much more complicated than often suggested. Your decision to nap will depend on many factors, including individual variation regarding the value and necessity of napping, the nature of your job, your chronotype, and whether you are struggling with insomnia. If you are a truck driver, a nuclear energy engineer, or an airline pilot and you are feeling sleepy on the job—or if you have a crucial task coming up that requires you to be sharp for a particular amount of time—please, take a nap. But if you have been having trouble sleeping at night—and this is a sizable hunk of the population—and you are simply feeling the urge to nap, it may not be the best move.

If you *are* going to nap, you want to do it correctly. And that is mostly about length and timing. "The briefer and earlier in the day, the better," is the rule of thumb, Professor Carney said. Studies have consistently found a varied response to the length of a nap. A short nap can be good, mid-length naps have a neutral health impact, and longer naps are associated with negative health outcomes. Also, short naps usually result in less post-nap sleep inertia. Most studies put the sweet spot between 10 and 20 minutes. (Unless, that is, you are napping for safety reasons, in which case you want to nap until you are no longer sleepy.)

The sleep portion of my napercize session lasted about 20 minutes. I was out like a light. And I gotta say, it felt darn good! But I also like cupcakes.

3:00 PM—Five-second rule

In our house, my wife's chocolate cake with seafoam icing is considered a near-spiritual experience. Any excuse to consume it is welcome. Once, when our kids were still in elementary school, we went to a neighbor's house with one of these magical cakes in hand. The door opened and there stood our friend's two exceptionally well-behaved children, dressed perfectly, every hair in place. All at once our unruly brood swelled forward, causing the cake to flip into the air. When it hit the floor it exploded like a chocolate grenade. Eyes wide and mouths open, our kids froze for a millisecond. With near superhuman speed, all four silently calculated their options. And then, in perfect unison, they pounced on the cake like a pack of hyenas. They were on their hands and knees inhaling cake, icing, and, probably, bits of plate. They likely figured they had mere seconds before the parental units would bring the feast to an end.

They had five seconds, to be exact.

Every parent knows this parenting truism: if a food item hits the floor, your child can still consume it if you pick it up fast enough. Apparently, our children had learned this lesson well.

Of course, the five-second rule isn't just for kids. Many of us believe it to be a scientific fact, one that is called upon by millions of people every day. According to a 2014 survey done at Aston University in the U.K., 87 percent of respondents admitted that they would or already had eaten food off the floor, and 81 percent said that, when considering the option, they apply the five-second rule.

What is your own floor-food policy? If you drop your afternoon doughnut on the office floor, is it safe to quickly retrieve and eat it?

What makes this decision interesting is the degree to which the five-second rule has attracted the veneer of scientific legitimacy. I've asked many people about the veracity of this rule. Many thought it was real, including those in the scientific community. And they often rationalized their conclusion by saying something like "I recall hearing about a scientific study on this." This "look, it's science" approach is often echoed in the media. And the fact that "five-second rule" is a clever catchphrase has also given the concept pop culture legs.

Like many other topics covered in this book, the five-second rule has become accepted wisdom because it feels right, it is easy to remember, and it is a fun idea that, if true, would be tremendously convenient. Who doesn't want it to be okay to pick up a doughnut from the office floor?

In fact, the science is pretty thin. The studies that are most frequently used to support it are a 2014 undergraduate student project (the same group that did the above-noted survey) that was not published in a peer-reviewed journal and a 2004 study conducted by a high school student that, again, was not published.

As far as I can tell, the most rigorous, peer-reviewed work on this topic is a study conducted in 2016 and published in the journal *Applied and Environmental Microbiology.* The researchers examined the bacteria transferred from floor to food using a variety of surfaces (stainless steel, ceramic tile, wood, and carpet) and different foods (watermelon, bread, bread and butter, and gummy candy). They contaminated the test floor with a nonpathogenic relative of the salmonella bacterium and left the food on the floor for varying periods of time.

What did they find? Basically, the five-second rule is a significant oversimplification of what actually happens with bacteria transfer from floor to food. In fact, all of the variables come into play, especially how much water the food contains. In this experiment, for example, the watermelon was contaminated with bacteria the quickest and most completely because moisture draws the germs onto the food. More importantly, the study found that for all the foods they used, some transfer takes place within less than a second of the food hitting the floor. As a result, the authors explicitly declare that their research disproves the five-second rule.

This conclusion accords with one of the only other peer-reviewed papers on the subject, a 2007 study published in the *Journal of Applied Microbiology*. That study also found that under the right conditions, bacteria "can be transferred to the foods tested almost immediately on contact."

To be fair to the five-second mythology, these more rigorous studies did find that if the floor is dry and hard and the food isn't particularly wet, the risk of transfer is smaller. But there really isn't a "rule" that can be applied to such a wide range of situations. Bottom line: bacteria can't count.

I asked the senior author of the peer-reviewed 2016 study, Professor Donald Schaffner at Rutgers University, why people are so quick to embrace this vastly oversimplified version of the science. "That's hard to say," he told me. "I think people are often looking for quick rules of thumb that justify what they want to do anyway." For Schaffner, who is a microbiologist and food safety expert, part of the problem is that the state of the knowledge prior to the peer-reviewed work was largely shaped by the media reporting on less than robust research that was nothing more than, to quote Schaffner, "publication by press release." I agree with Schaffner. This seemingly trivial topic is a wonderful example of how an appealing how-to-live-your-life catchphrase can be embraced by both pop culture and the public, which then leads to questionable research that leverages that interest—which, in turn, leads to media portrayals of that somewhat less than rigorous research.

Given the ubiquity of this topic, I thought it best to double-check with yet another academic expert. I turned to Professor Brett Finlay, the well-known germ expert from the University of British Columbia we met earlier in this book. He agrees that the science has been oversimplified. For Finlay, it is more about the surface that your food hits than the time it spends on the ground. "Hard synthetic surfaces are good. Soft gooey surfaces are not. Time doesn't really matter," he told me.

This more nuanced conclusion should surprise absolutely no one. Of course it depends on the type of food involved, the nature of the floor surface, and where that floor is located! If your doughnut lands on the floor in a hospital, fast-food kitchen, or sewage processing plant, the amount of time it is on the floor is irrelevant. Don't eat the doughnut. Repeat: bacteria can't count.

If, however, your birthday cake lands on a neighbor's dry, spotless hardwood floor, I say go for it.

It is fair to say that Professor Schaffner has emerged as the world's leading expert on the five-second rule, and so I couldn't resist asking him if *he* would eat the office-floor doughnut. "One of the important questions that one has to ask is, How likely is the surface that my food contacted to be contaminated?" he mused. "I might judge a New York City subway platform differently than the floor of my office." While he figures a doughnut is a fairly dry item—and therefore less likely to suck up bacteria—in the end it comes down to assessing the surface on which the doughnut in question resides. "How clean do you think your office floor might be? Who's been walking on the floor, and where else might they have walked?" And who trusts the bottom of their office mates' footwear?

But even this renowned microbiologist will implement a cost/benefit analysis when deciding what to pick up off the floor. "My rule of thumb is that I am pretty circumspect when it comes to food dropped in places that are outside my house," Schaffner said. "Even in my house, I would never eat a piece of wet food that fell on the floor, but if it's a yummy chocolate chip cookie, sure."

I get that. A yummy chocolate chip cookie will beat the risk of a potentially pathogenic germ every time.

3:15 PM—Email

As the end of the day approaches, I start to get panicky about all the unanswered email hanging out in my inbox. I keep opening my inbox and clicking on messages that I think I can handle quickly. Often I end up doing nothing more than opening and closing messages without making any real progress. Open, scan, close. Open, scan, close. And while I'm doing this, 10 more emails arrive. Argh!

I'm not alone in my email angst. The average person has 199 unread emails in their inbox. And many of us, myself included, feel a strange compulsion to get our inboxes down to zero, though 24 percent of us think "inbox zero," as it is lovingly called, is a fantasy-like dreamland. Given that the average inbox—at least according to one industry analysis of over 38,000 of them—contains more than 8,000 messages, and about 20 percent of us have more than 20,000, the inbox zero goal does seem like a bit of wishful thinking.

Computer programmer Raymond Tomlinson sent the first email, in 1971. (That message was "QWERTYUIOP.") Today, the average number of emails sent each day hovers around 270 billion.

If you work in an office, email likely devours much of your life. A 2017 study from Carleton University found that email consumes one-third of our time at work. When people work at home, about half the time is spent reading and answering email. Another study, a survey of 1,000 white-collar workers in the U.S., calculates that the time suck is even more significant, finding that the average person spends 4.1 hours a day on email. Think about these numbers for a moment. For many of us, our jobs are, basically, about managing email.

Most of us have a love-hate relationship with email. On the one hand, studies have found that it remains the preferred method of work communication. A 2018 industry analysis found that "86 percent of professionals prefer to use email when communicating for business purposes." This is certainly the case for me. I view my inbox as a "to-do list" to be aggressively tackled every single day (which, as we will see, is probably a bad idea). I don't need or want other modes of contact or communication. I ignore my office phone when it rings and have a message that states I will not reply to voice mail. (Of course, most callers still leave a message.) I rarely answer the landline at home (which drives my family nuts). I leave my cellphone off (which drives my family nuts). And I generally ignore Twitter, Instagram, and Facebook direct messaging. For me, and for so many others, email is king.

On the other hand, a growing body of research has found a correlation between large email volume and higher levels of work stress and lower levels of work satisfaction. A 2011 study, for example, found that "the more time people spent handling e-mail, the greater was their sense of being overloaded." A 2016 study of almost 400 office workers concluded that "both the actual time spent on emails and organizational expectations regarding employee availability to monitor work emails after hours lead to emotional exhaustion." Biomedical investigations have demonstrated that email causes a significant rise in our heart rate, blood pressure, and cortisol levels (cortisol is the primary stress hormone). Stress levels are highest when inboxes are the fullest. Being interrupted by email can also dramatically affect productivity. According to 2012 research from the U.K., "A typical task takes one third longer than undertaking a task with no e-mail interruptions." And several studies have found that email stress follows us home. We get stressed just thinking about email!

So yes, we all recognize that we need email, but we all also absolutely hate it. In an article on the history of email, tech writer John Pavlus called email "the most reviled communication experience ever." I am squarely in this camp. Email feels like an evil, time-sucking, joy-killing grey blob that follows me everywhere I go.

(By the way, we especially hate particular kinds of email. A 2018 survey found that the most annoying phrase to find in your email is "Not sure if you saw my last email." Yep, I saw your last email, ignored it, and now you are sending me another email that I will also need to ignore.)

Even though most of us recognize that email has a detrimental effect on our work and home life, we find it hard to escape its grasp. Studies have found that about one-third of us wake up in the middle of the night to check our email, and 79 percent of us check our work email while on vacation (that's me), with one-quarter checking it frequently (also me). And when we do look at our email, we look at it almost as soon as it pops up on our screen. One analysis found that 70 percent of emails were opened within six seconds of their arrival and 85 percent were reacted to in some way within two minutes.

We usually try to answer pretty quickly, too. A survey of 503 U.S. employees found that about one-third answer emails within 15 minutes. Another 23 percent say they answer emails within 30 minutes. The survey also found that many people admitted checking their work email during funerals, weddings, and even while their wife was in labor. (I have four kids. I might have glanced at my inbox during the birth of the fourth, because, well, he was the fourth and things seemed to be going well.) A 2015 study of over two million email users concluded that as email volume increases, our email responses become shorter. No surprise there. But the researchers also discovered that despite the increased volume, responsiveness may even increase. Like a rat on a wheel, we scramble to keep up—but we rarely manage to.

What can we do? Email triggers a cascade of dozens and, for some of us, hundreds of decisions each day. How we choose to handle these decisions can significantly affect both our work environment and our well-being. Like so many things covered in this book, there is no magic answer, even though countless business gurus and wellness pontificators profess to have a solution to our email angst. The reasons we are stressed out by email are shaped by many factors, including the type of work we do, our personalities, the nature of our home life, and, perhaps most important, our workplace culture and expectations. It is unrealistic to presume that a one-size-fits-all fix exists for this socially complex phenomenon.

Still, looking at both the available evidence and the realities of the work environment (most of us can't, as much as we would like to, ignore our email), we can see a few broad rules that can help mediate the anxiety.

How often should you check your email? A 2017 analysis of 225 million hours of work time found that, on average, we check email about once every 7.5 minutes. That is probably way too often. But the answer to this question is confused by an emerging (and overly simplistic) time management mythology focused on avoiding the inbox for long stretches of time. A common recommendation permeating the popular press is that we should all be batching our email and checking it just a few times a day. The idea is that this will allow us to stay focused and avoid wasting time checking email as it comes in, thus reducing stress and helping us avoid the loss of concentration and productivity associated with being distracted. There is some empirical research to support this approach. A 2015 study from the University of British Columbia randomly assigned 124 people to a week of checking their email three times a day or checking an unlimited number of times. When that week was up, everyone switched strategies. What the study found was that during the week of restricted email checking, the participants experienced significantly lower stress and reported higher levels of general well-being.

Although such research is compelling and the "check less" concept feels instinctively correct, it was just one study of mostly university students, who hardly face the email expectations of a typical office employee. Indeed, people were excluded from the study unless they "had some flexibility in how often they could check their email and were interested in experimenting with the way they managed their email." In other words, the study selected for people who aren't ruled by their inbox, making the results far from generalizable. When you are a student, you can look at your email when you want. When you work for a large law firm or a financial institution or a tech company or a hospital or the media, it might be more difficult to take a Zen approach to your inbox. In fact, as we will see, ignoring your email might cause more stress!

Still, this study generated a great deal of media interest and is often used as the rationale for the now pervasive "less checking/more batching" approach. Do a web search for "checking email less frequently reduces stress" and you will get pages and pages of articles, blogs, and news reports referencing this one piece of research. It is the kind of message that feeds into the business guru advice industry ("The Seven Email Habits of Successful People!") and quickly gains cultural momentum.

To be clear, I think this is a valuable, well-done study that informs our email management strategies. But, as we have already seen in other situations, the underlying evidence is often more complex than how it's represented in the public sphere. For example, a 2016 study of office workers employed in large corporations found that batching email did result in a perceived increase in productivity, but, the authors wrote, "despite widespread claims, we found no evidence that batching email leads to lower stress." Another small study, published in 2017, explored the psychological impact of removing all electronic communication notifications. While the participants felt more productive at work, the lack of contact and the inability "to be as responsive as expected" made them feel more anxious. And they felt less connected to their social groups. The paradox created by limiting email access is captured in the title of this study: "Productive, Anxious, Lonely."

So what is the answer? I think successful email management is like successful dieting: the only way to succeed is to find a strategy that you can actually maintain, one that works for you and your work environment. Everyone—and every workplace—is going to have different expectations and needs. That said, whatever strategy you adopt, it should nudge you toward establishing hunks of email-free time and, if practical in your work, turning off email alerts. (Despite what is often recommended in the popular press, this may mean aiming for relatively short bursts of email-free work time.)

A 2017 review of the literature, authored by Professor Emma Russell of Kingston University in London, came to a similar conclusion. She determined that it was a myth that we should check our email only a few times a day. In fact, she says we should "check and process email regularly in order to prioritise and control our work effectively." To do otherwise can lead to anxiety about an overflowing inbox.

I asked Russell what she thought of the batching idea. Email, she told me, "has become so much more multi-faceted and functional that it can't be 'parked' as a sideshow anymore, but needs to be integrated fully into our working lives. It seems easy to say 'leave it' and deal with it just two or three times a day. This is appealing. But our research has found that people who do not integrate email into their working day will experience big build-ups in the inbox and that this can cause overload, stress and a sense of loss of control."

Professor Russell's review recommends turning off alerts and "checking, and then deleting, filing or actioning, email every 45 minutes" in order to "tangibly reduce stress and improve efficiency." This strikes me as practical advice. And Russell told me that if they adhere to these rules, most people do become more efficient. Pick an email-free interval that fits your lifestyle, temperament, and workplace expectations. It might be 45 minutes or it might be two hours.

When should you answer your email? Professor Dan Ariely, a behavioral economist at Duke University, found that, for most of us, only 12 percent of email could be categorized as urgent and worthy of a quick response. And over one-third never needs to be answered at all—simply delete it. Much of the other email can be either temporarily ignored or filed away. One small study suggests that the best approach for filing is to have only a few folders, to avoid overthinking where to place an email. (But let's be honest, you will probably never look at it again!) This means that over a workday there are probably only a few emails that really require quick action. Concentrate on answering those few and relax about the rest. After all, someone is also likely ignoring your emails.

In fact, another source of email angst is the expectation that people will answer *your* email quickly. We live in a world where people carry their phones nearly constantly and check them nonstop. So it is easy to assume that the recipient of your message will see and respond to *your* insightful email lickety-split, even though you are simultaneously disregarding someone else's. Workplace surveys have found that most people expect their email to be answered within a day and often much sooner. And the expectation for a rapid response is accelerating, resulting in the need for a generous give-and-take email ethos. If we are going to relax about checking email, we also need to relax about expecting a response. Do not, unless absolutely necessary, be the author of an email that begins, "Not sure if you saw my last email."

If you really do want a fast response, what time of day is the best for sending? An analysis by an online marketing firm studied over 500,000 emails to get a sense of when they were opened and replied to. Their conclusion? The best time to send an email is on the weekend, early in the morning or early evening, because reply rates were about 6 percent higher on the weekend than on weekdays—email traffic is lower and you will have less email competition. But this kind of thinking is what has allowed emails to creep into every corner of our lives. Indeed, this advice is a tad depressing. It is premised on the idea—backed by data—that we should work on the weekend and that we can assume that other people will be working on the weekend too. And because this is how the world now operates, we should exploit this sad reality. I'm not going to recommend you act on such life-destroying advice (though don't be surprised if you receive an email from me at seven one Saturday morning).

How should we answer? We find emails stressful for many reasons, but a big one is that they ask us to do something. In a 2018 article in the *Harvard Business Review*, author and business professor Dorie Clark described how often her own inbox was a source of appeals for her time and energy. "I received 69 requests per week, or nearly 10 per day. It takes extreme willpower to say no, but it became easier when taking the aggregate numbers into account."

For me, Clark's last point is key. Saying no is stressful. And this is a big reason why I find my email inbox stressful. It is a list of appeals for my time. The guilt associated with sending a friend or colleague a no weighs heavily. But it is impossible to say yes to everything. Formulating strategies for saying no can help make your inbox less anxiety-provoking.

Online you can find plenty of business-speak recommendations on how to say no (you must synergize your blue-sky thinking in order to optimize a multi-platform refusal), but there isn't much actual empirical research on exactly how to do this. One of the few academics doing research on how best to tell the world no is Professor Vanessa Patrick at the University of Houston. She has explored, for example, how different ways of saying no ("I can't" vs. "I don't") allow people to stay committed to their refusal. ("I don't" works best, in case you are wondering.)

"One of my pet peeves is people speaking about this without any understanding or research," Professor Patrick said when I asked her about the ubiquitous online advice on how best to say no. This advice is usually not grounded on evidence, she told me. Like the recommendation to batch email, the "just say no" advice is often oversimplified because it ignores individual circumstances. What is needed, Patrick suggests, is a strategy that fits the person. Her current research is exploring how to do just that. She has found that one of the reasons people get anxious responding to requests, whether in person or in email, is what she calls "the spotlight of refusal." Basically, the focus is on us to take some action. This can feel stressful. We want the spotlight to dissipate. "A good refusal needs to do two things," Patrick said. "It needs to convey a clear message—that you are saying no—but still maintain the relationship."

Patrick said her research supports a modified version of the classic "it's not you, it's me" breakup strategy. "Figure out your priorities.

Develop a personal policy"—such as, I'm not taking on more talks, I don't work on weekends. "And then explain this is how you operate. It's not about them. It's about you," Patrick said. You will feel less anxious because you have a plan of action that is tailored to your goals and personality. And, Patrick explained, the receiver of your negative response will, in general, respect your decision. Research Patrick published in 2012 found that "invoking a personal policy conveys greater conviction and results in less push-back."

As email traffic continues to climb in our working lives, understanding how we manage requests for our time is increasingly essential. Patrick and her team are working on this too. They are, for example, exploring how emojis might be used to convey a firm but kind refusal. "Sending an emoji can be effective—a nonverbal way to say no," Patrick told me. "It can add humor and act as a buffer. A softer no."

This is exactly what I need—one emoji that conveys the following message: "My schedule is full, your request isn't a priority for me, and, by the way, it's not you, it's me." Send.

So what are my big email takeaways? Of course, much of the problem stems from workplace culture. Given the costs of email expectations to our well-being, work efficiency, and family life, it would be most helpful if companies and institutions recognized that change is needed. And in some places, this is happening. In France in 2017, for example, a new law established workers' "right to disconnect." If a company has more than 50 employees, it must create an after-hours time period when staff are not to send or answer email. A similar law was proposed for New York City in 2018. And some companies, such as Volkswagen in Germany, turn off their email servers late at night to "respect relaxation time."

Despite these policy developments, for most of us, email expectations have not abated. And most of us aren't in a position to change our workplace email demands. What should we do? In keeping with the theme of this book, when it comes to our email, we should do our best to just relax and find what works for us. Email is clearly a source of significant stress, but adopting some trendy management scheme to get your inbox down to zero could end up being a waste of time and psychic energy if it doesn't help you feel less anxious and allow you to accomplish real, productive work. As Professor Ariely has suggested, focusing on perfect email management can amount to little more than "structured procrastination." Indeed, no one lies on their deathbed thinking, "I wish I'd spent more time organizing my email." Likewise, batching and ignoring email—though intuitively appealing—is probably an unrealistic and potentially stress-inducing option.

My suggestion is to, yep, relax. Keep it simple and incorporate a couple of broad general themes: turn off your alerts, do your best to create a few blocks of email-free time, focus on the key emails, file away or delete as much as possible, and develop a strategy to simply say no.

I'm still working on that last one.

4:00 PM—Handshake

According to a 2018 headline from the *New York Post*, research has uncovered that "people with a strong handshake are more intelligent." If true, I'm in big trouble. I have a terrible handshake. I always get the timing wrong and end up leaving a few fingers uninvolved, thus creating what a handshake aficionado might call "the dead fish." Recently I've started asking for a do-over. "Let's try that greeting again," I'll say. Usually the recipient will immediately agree, which I take as a silent recognition that my first attempt was clearly substandard.

Although the exact origins of the handshake are unclear, the practice has been around at least a few thousand years. A ninth-century BC stone carving depicts the Assyrian king shaking it up with a Babylonian ruler. It has been said that the handshake is a symbol of peace, showing that neither party is carrying a weapon, but a 2015 study has suggested that the gesture may have more biologically oriented roots. The research, which involved covertly videotaping hundreds of people shaking hands, found that participants increased the sniffing of their hand by more than 100 percent, suggesting that handshaking may have started as a way to sample a stranger's odor (which is, I suppose, better than the approach adopted by dogs). But regardless of its history and purpose, the handshake has become the most ubiquitous form of greeting throughout the world and the recognized standard greeting in the work environment, at least prior to the coronavirus outbreak of 2020.

There is a bit of evidence that a solid handshake may help you make a good first impression. A small 2012 study, for example, used brain scans to explore whether a handshake created a positive interaction. It found, to quote the study's authors, that "a handshake preceding social interactions positively influenced the way individuals evaluated the social interaction partners and their interest in further interactions, while reversing the impact of negative impressions." In other words, a handshake helps to make a good impression and can make a bad impression a little less bad. A series of studies published in 2018 by researchers at the University of California, Berkeley, indicate that handshakes "signal cooperative intent and promote deal-making." The lead author of the study, Professor Juliana Schroeder, said that handshakes make constructive cooperation more likely because they change the dynamic of a situation. "It changes the way you perceive not just the other person, but the way you frame the whole game," Schroeder said. "You say to yourself, 'Now we are in a cooperative setting rather than an antagonistic one.'"

But before you shake, you may want to consider the issue of germs. As we saw earlier, a large percentage of the population doesn't do a great job of washing their hands. It has been estimated that, on average, people will shake about 15,000 hands in their lifetime—which is a lot of dirty hands.

It's no surprise, then, that handshakes have been identified as one way pathogenic germs get transmitted from person to person. With this in mind, four infectious disease experts from UCLA, in a 2014 article in the *Journal of the American Medical Association*, recommend banning the handshake in health care settings. But these experts also recognize that avoiding handshakes comes at a cost. "In an attempt to avoid contracting or spreading infection," they write, "many individuals have made their own efforts to avoid shaking hands in various settings but, in doing so, may face social, political, and even financial risks." As such, they figure we need an acceptable replacement, one that may provide some of the same social benefits as a handshake but with less of the physical contact that allows germ transmission. The answer: the fist bump.

A similar no-handshake strategy has been proposed for cruise ships. There have been many high-profile germ-induced cruise ship outbreaks—most commonly associated with the rabid spread of gastroenteritis from norovirus and the novel coronavirus COVID-19. Because cruise ships are self-contained communities, they have the potential to become floating incubators for infectious diseases. Consequently, it has been suggested that passengers use only the "cruise tap" to greet their fellow cruisers. The cruise tap is a refined version of the fist bump, whereby only two knuckles briefly touch. This is next-level germ avoidance.

Several studies support the idea that the fist bump—and, by extension, the cruise tap—conveys far fewer germs than a handshake. A 2014 study from Aberystwyth University in Wales explored the germiness of various greetings. What it found was that handshakes transmit twice as much bacteria as high fives and 10 times more bacteria than fist bumps. So if germs are the concern, you may want to go with the fist bump, particularly during flu season.

But how concerned do we really need to be about germs in this context? If you work in a hospital or you're on a cruise ship, the risks associated with handshakes are real. The stakes are higher than in most handshake situations, such as an office meeting or saying goodnight to a colleague. In fact, there is surprisingly little research on the germ transmission of handshakes outside of the health care setting. One study from 2011 measured the rate of contamination caused by handshakes during graduation ceremonies. It recruited people officiating at university, high school, and elementary events—people who shook a lot of hands in a very short period of time. What the researchers found after analyzing literally thousands of handshakes was that a lot of germs were passed around, but only rarely were those germs pathogenic. In fact, they found that the risk is one pathogen germ transmission for every 5,209 handshakes. That is a very small risk.

This study has limits—for example, graduation handshakes can be pretty pro forma and brief—but it is still a good reminder to relax. Not every handshake is going to lead to an Ebola outbreak.

What about that study that suggested my poor handshake meant I wasn't terribly bright? The study that generated the headline—and there were many similar headlines all over the world—was an observational study (remember, be careful with observational studies, which suggest only a correlation exists) about the relationship between grip strength and cognitive functioning (such as memory, recall, reasoning) in the general public. This relationship likely exists because grip strength is simply a marker for fitness: if you have good grip strength, you probably have more muscle tone. But the study had nothing to do with the quality of handshakes. This is, yet again, a case of the research being twisted in the service of a sexy headline. Sometimes a bad handshake is just a bad handshake.

So, to shake or not to shake? Given the deep history and cultural complexity of the handshake, it will likely take a bit of time to change human behavior, especially since the handshake does have potential benefits. And it can be hard to ignore a welcoming, outstretched hand.

There are situations, though, where it is reasonable to avoid handshakes, such as in health care facilities or if you are ill with some infectious ailment. Personally, I'm a fan of the fist bump (the full one, not that wimpy cruise tap), and I think society should segue toward this less germy greeting. Still, active pandemics aside, of course, we can all probably relax about the occasional handshake. A little human contact can have real benefits. And this brings me to hugging.

4:00 PM—Hug

In addition to struggling with handshakes, I am not a hugger. Yes, I frequently hug my wife, kids, and cat—especially the cat. But I hug my brothers, with whom I am extremely close and love almost as much as my cat, about once a year, perhaps less. And that is the full list of the biological entities that I want to hug. All the other hugging I partake in is either somewhat or entirely coerced by social circumstances.

Unfortunately for me and the other non-huggers out there, these are challenging times, hug-wise. The hug has infiltrated almost every social situation, including the work environment. Where once a handshake would suffice, now a hug seems compulsory. People I barely know will approach me with arms opened wide. Once a store clerk gave me a hug after I bought a pair of jeans. (These were the overpriced raw denim variety, so perhaps she felt sorry for me due to my bad judgment.) How can you refuse a hug invitation without looking like an unsociable jerk? You can't. And, as a result, an awkward, where-do-I-put-my-hands moment inevitably ensues.

When I'm on my toes, I'll try to preempt these situations with a quick and clear arm-out-for-a-handshake gesture, letting all parties know that

I view this as a handshake moment. While this strategy often works well, it is a high-risk move. The approaching acquaintance may start a hug attempt at the exact same instant as my handshake demand, thus recreating a kind of rock-paper-scissors referendum on the status of our relationship. As noted, these are precarious times for us non-huggers.

Now, I'm not saying that my position on hugging is correct. In fact, a small body of research does suggest hugging is good for us. A 2015 study published in *Psychological Science* found that people who hug are less likely to come down with a cold. The researchers speculate that hugging increased the perception of social support and, as a result, had a stress-buffering effect that helps the immune response. The same study found that the frequent huggees who did get infected felt less ill compared with the no-hug participants. Another study, published in 2018, followed more than 400 adults for two weeks and found that hugging was associated with a better reaction to social conflict.

This research is certainly interesting. And, like many of the topics in this book, the findings accord with our intuition about how the world works; specifically, the idea that human contact is valuable. No surprise, both of these studies led to a ridiculous number of stories in the popular press about the health benefits of hugging. But despite all the hugging noise, we need to remember that there are only a few relevant studies and that they are only correlational in nature. They don't *prove* that hugging leads to better health. Teasing out the effects of hugging on health would be a methodological challenge. Many variables need to be considered: What kind of hugging? Do shoulder-bump man hugs count? How long do you need to hug? What kinds of illnesses are you trying to avoid? Etcetera, etcetera. Indeed, one small study from Japan raises the intriguing question of whether you even need to be hugged by a human to derive the alleged benefits. The researchers found that being hugged by a robot led to a positive emotional response. Bring on the eHug.

Don't get me wrong—relationships and human contact matter, deeply. A growing body of literature supports the health value of good social relationships. And being touched by someone with whom you have an emotional bond can have a remarkable and measurable effect on your well-being.

But that doesn't mean that embracing your work colleagues will necessarily make you healthier. Hugging Stan from accounting at the end of a budget meeting is fine, if you and Stan want to hug it out. But don't feel pressured to hug Stan if that's not your thing. My vote: let's either cool the hugs-for-everyone trend, or workplaces should buy some hugging robots to take the heat for the non-huggers like me.

4:30 PM—Time panic!

As Dr. Seuss said: "How did it get so late so soon?"

The end of the workday always feels to me like it arrives with no warning. I've accomplished one-tenth of what I wanted to do. All of that unfinished stuff is plowed into the next day. This means that tomorrow I'll finish only one-twentieth of what I hope to finish. And on it goes.

Almost everyone feels this way. One industry survey reports that 58 percent of Americans would pay $2,700 for an extra hour in their day. According to another study, published in the *Proceedings of the National Academy of Sciences*, "Working adults report greater happiness after spending money on a time-saving purchase than on a material purchase." Don't spend your money on stuff. Use it to buy free time.

There simply isn't enough time in the day. But is this actually true, or is this perception of a time famine, as it is often called, an illusion?

Although we may feel ridiculously busy, many of us don't actually spend that much time at work doing actual work. According to a study from the U.K. involving almost 2,000 office workers, the average person is productive just 2 hours and 53 minutes a day. The rest of the time is spent on such things as social media, reading the news, socializing, and eating. This accords with a 2017 study, noted above in the email section, which monitored workers' computer activity during 225,000 million hours of work time and found that most of us are productive only 12.5 hours in a week. In another survey of 3,000 workers, published in 2018, 45 percent of employees worldwide said they could do their jobs in five hours a day or less, even though they actually worked in full-time positions. That same survey found that 71 percent believe work interferes with their personal life.

Thus an odd time paradox exists. Many people simultaneously feel far too busy and also that they waste a significant amount of their time while at work. In fact, they are right on both counts.

While we may think humans have never been busier, the evidence suggests that this likely isn't true. Research shows that we often overestimate the amount of time we spend working. And the busier we think we are, the more we overestimate. An analysis done by the University of Oxford's Centre for Time Use Research of data from approximately 850,000 time-use diaries covering a span of 50 years found that workers who claim to work 75 hours a week often overestimate by as much as 50 percent. And some professions are particularly bad, among them lawyers, teachers, and police officers, who overestimate by more than 20 percent.

Most studies show that the time we spend at work has *decreased* at a pretty steady pace over the past few decades, a conclusion confirmed by the Oxford Centre's study, which found that leisure time has increased for most individuals. Yes, certain groups, such as single parents, have extremely busy lives. In addition, I realize that many individuals, because of various life circumstances, work a crazy amount of hours. And there are significant differences between various sectors of the economy. There are professions, including law, finance, and consulting, that glorify and reward long hours. (Research suggests that this has contributed to the gender pay gap, because men are more likely to engage in overwork.) But in the aggregate, we work less now than at any time over the past century or so—at least in wealthy countries. In 1870, for example, working 60 or 70 hours a week was the norm, not the work-martyr exception. Data from the OECD (which stands for the Organization of Economic Cooperation and Development and is, basically, the club of the more affluent nations) shows that in the year 2000 the average full-time worker put in 1,841 hours a year; by 2017 that number had dipped to 1,759. In the U.S., the hours went from 1,832 to 1,780 and in Canada from 1,779 to 1,695. Any way you slice it, we are working less.

And let's not forget that the average person spends a lot of time doing stuff that isn't usually thought of as "being busy." Most people in the U.S. spend almost three hours a day watching TV. People spend more time watching the tube than on childcare, exercise, cooking, or household chores. A survey done by Deloitte in 2018 found that Americans watch about 38 hours of video a week, which is basically the equivalent of a full-time job. The next most common nonemployment activity is socializing and communicating with actual humans, which accounts for just 39 minutes a day. And here is another sad stat: people aged 15 to 44 read for just 10 minutes a day. People may *feel* busy, but they find time to watch TV for three hours a day. I'm not judging. Really. There is a lot of great TV out there. But does watching it qualify as being "busy"?

Something that may have contributed to the "we're all too busy" perception is the trend of "busy bragging." In the recent past, having lots of leisure time was a sign of affluence. Wealthy, high-status people hung out on their estates, *Great Gatsby* style, doing mostly nothing beyond looking fabulous. Being idle was a sign of status. Now, being busy is the badge of honor. It signals that you are valued and doing important work.

A series of studies led by Silvia Bellezza of Columbia University and published in the *Journal of Consumer Research* in 2017 supports the existence of the busy-bragging phenomenon. In one study they assigned one group of people to read leisurely posts on Facebook and another group to read posts that suggested the writer was busy. They found that busy people were perceived as possessing more status than those who were not signaling that their schedule was packed. People appear to sense that their social status rises when they are perceived to be busy, and as a result they have a tendency to busy-brag.

Professor Bellezza is an expert on how consumers use products to express who they are. "In advanced economies," she told me, "long hours of work and busyness may operate as a signal that one possesses desirable human capital capabilities." A busy brag tells the world you are "in high demand and scarce in the job market, leading to elevated status attributions."

Professor Bellezza and her colleagues also examined how high-status individuals, particularly celebrities, busy-brag on social media. Of the 1,100 tweets they analyzed, 12 percent related to complaints about being too busy and lacking free time. "I have been so ridic busy w meetings and calls that I have neglected my fans" is a typical celebrity busy-brag. Such messaging helps to normalize both the compulsion to busy-brag and the idea that successful, high-status people are always busy—which fuels the perception that, if you think of yourself as successful, you should feel busy too.

Many of us have internalized this idea. When someone asks us how we are doing, our first response is often some version of "I'm okay, but I'm so darn busy!" (I *still* do this, even after studying the busy-brag phenomenon. It has become a norm. A new small-talk tic.) We feel this way in part because we feel we *should* be busy and, likely unconsciously, want the world to see us that way too. As a result, Professor Bellezza told me, "in contemporary American culture, complaining about being busy and working all the time has become an increasingly widespread phenomenon." By way of example, she highlighted an analysis of holiday letters that found that "references to 'crazy schedules' have dramatically increased since the 1960s." When we tell people about our year, we want the takeaway to be one clear message: We are all busy! The kids are busy, the cat and dog are busy, and, especially, the letter writer is busy. No one says, "We had a great year watching all six seasons of *House of Cards*. We even watched the one without Kevin Spacey! Crazy, I know. Like last year, Little Johnny spent about nine hours a day online. Such a typical teenager!"

The fact that we carry our office around in our pockets further contributes to our sense that we are too busy. There is always something we *could* be doing—checking email, calling colleagues, reviewing documents. This may make us feel that we are "busy" when in fact it is simply the vague sense that we *should* be working because we *could* be working. Indeed, there is some evidence that people seek to feel busy because to feel otherwise is to feel we are being irresponsibly unproductive. A review of the research surrounding how we perceive time, published in 2019 in *Current Opinion in Psychology*, speculates that people endeavor to work hard not to create free time but to avoid being idle, because being idle means we are not valued.

Finally, the nature of work has changed. In the past, many jobs had a concrete endpoint. You bring in the crops until the field is empty. You build items on an assembly line until you are done. But, as noted by journalist Oliver Burkeman, in our knowledge economy "there are *always* more incoming emails, more meetings, more things to read, more ideas to follow up—and digital mobile technology means you can easily crank through a few more to-do list items at home, or on holiday, or at the gym." The result, inevitably, is that we feel overwhelmed.

The claim that we are not as busy as we think and say we are may seem vaguely insulting, as if I'm both accusing people of lying and suggesting that many of us don't possess, as Professor Bellezza put it, "desirable human capital capabilities." But I think that many of us, deep down, know we aren't truly insanely busy. A 2015 study of over 10,000 people from 28 countries found that 42 percent admit to overstating how busy they are and 60 percent believe that most other people do this too. For millennials, the numbers are higher: 51 percent admit to lying about their busyness and 65 percent suspect that most other people do the same.

Many of us aren't experiencing a genuine time famine. It would be more accurate to call it a time paradox. Many factors, including a changing work environment and a desire to project a certain image, make us feel as if we are scrambling, when a more objective assessment suggests that this isn't really the case—or, at least, it doesn't have to be.

In 1930 the economist John Maynard Keynes made a famous prediction that, because of economic and technological improvements, his grandchildren would need to work only three hours a day, and only if they wanted to. If you believe the data on how many hours a day we are actually productive, Keynes wasn't that far off, at least in his prediction of hours devoted to actual work. Of course, he totally missed the impact technology would have on our leisure time. Rather than liberating us, it has made us *feel* like we are constantly busy, even when we aren't. (Again, I'm excluding certain categories: people in the professions where working more than 50 hours a week is encouraged, people on changing shift work schedules, and people who take on multiple jobs to make ends meet.)

Feeling busy can have both good and bad consequences for our personal and work lives. On the one hand, feeling too busy to the point of stress can have adverse effects on our health and our work environment. We become less productive and less creative when we work too long and feel overly busy. We may try to engage in counterproductive multitasking that can, ironically, make us feel even more time crunched. And feeling as if we are in a constant scramble can lead to poor sleep and bad health decisions.

On the other hand, there are upsides to a busy vibe. People who feel busy are more likely to finish tasks, and they take less time to do so, probably because they are motivated to use their time efficiently. Also, the right level of busyness—that sweet spot where we feel we have lots on the go but aren't in a panicky kerfuffle—is associated with increased happiness. There is some data (correlational in nature) that suggests busier cities are happier cities. And a busy mindset is associated with an ability to exert self-control. A 2018 study from Hong Kong, for example, explored how "a busy mind-set bolsters people's sense of self-importance, which, in turn, can increase self-control." And this self-control can lead to potentially healthier decisions about things like diet and exercise. As one of the authors of the study, Amitava Chattopadhyay, summarized in a press release about the research, "When we perceive ourselves to be busy, it boosts our self-esteem, tipping the balance in favor of the more virtuous choice."

In addition, there is a bit of evidence that being busy is associated with improved cognitive abilities. A 2016 study of over 300 American adults "revealed that greater busyness was associated with better processing speed, working memory, episodic memory, reasoning, and crystallized knowledge."

As always, we need to be careful not to overinterpret these busyness studies—busyness is a hard phenomenon to study well. Much of the research is correlational—that is, being busy is merely associated with a particular outcome. We don't know if it caused the outcome. And what, exactly, is "busyness"? Do we all experience it the same way? Is it a discrete, consistent variable that can be meaningfully and reliably measured? Despite the methodological challenges of studying the phenomenon, it seems safe to conclude that the concept of busyness is a bit slipperier than is often portrayed. It is neither an affliction that needs to be eradicated, nor a state of being that should be revered.

Still, many of us say that busyness has an adverse effect on our lives. As noted in a review of the literature published in 2019, "Myriad surveys have recently found that approximately two thirds of Americans say that they always or sometimes feel rushed and half say that they almost never feel that they have time on their hands."

—

What can be done? There are, thankfully, many practical strategies that we can use to reduce busyness-associated stress, including the straightforward suggestion to just slow down. While this may seem both overly simplistic and contradictory (how can slowing down help you get things done?), the evidence shows that taking a moment to breathe (literally breathe) and to focus on a single task can moderate the anxiety associated with busyness. Focus on what needs to get done. A related strategy—already touched on several times in this hypothetical day—is to avoid multitasking. As much as practical within the bounds of your job, do your best to concentrate on one task at a time. This not only increases productivity but also helps to reduce that sense of fragmentation that contributes to busyness-related stress.

Another bit of sensible advice is to try to create a discrete break between your work life and your family and leisure time. Just because you have your office in your pocket doesn't mean you need to be on call 24 hours a day. A lack of "boundary management," as it is often called in the scientific literature, is associated with increased stress and a decreased sense of well-being. For example, a 2018 study of almost 2,000 employees found that those who "scored high on work-to-life integration enactment"—that is, less boundary management—"reported less recovery activities and in turn were more exhausted and experienced less work-life balance." Creating boundaries helps to stop work and busyness stress from seeping into every corner of your life.

You should also recognize that you aren't going to get everything done. You can't win the do-everything-now battle. Just accept it. Time will always win. For me, accepting this reality is liberating, as it forces me to concentrate my energies on the stuff that deserves my time. And remember to consider Hofstadter's law when estimating the amount of time it will take to tackle your priorities: "It always takes longer than you expect, even when you take into account Hofstadter's law." Studies have shown that we use overly optimistic predictions of how quickly we can complete a task. (This is called the "planning fallacy"—one example, of many, of how we humans are often overly optimistic about our own capabilities.) Interestingly, we are more pessimistic about how fast other people will complete their tasks. We think: "I'll do this quickly and efficiently. My colleague Stan will take forever." But nope, you are probably as slow as Stan. Plan for that.

For me, the best recommendation is one that is the most direct. This is a book about decisions, so my advice is to *decide* not to buy into the busyness bullshit. You don't win points on your deathbed for being the person who felt the busiest. An honest assessment of your actual schedule and a reconceptualization of your sense of busyness can help to give you a more relaxed perspective. Easier said than done and too simplistic to work, you say? Remember that our busyness stress is, to a large extent, a perception issue created by cultural and technological pressures. It makes sense that rejigging that perception—that subjective assessment of time—can have a beneficial result.

Some empirical evidence does support this strategy. A 2015 study published in the *Journal of Marketing Research* involved a series of experiments designed to either generate or reduce the stress associated with feeling busy. What it found is that if people thought about their time in the context of goal conflicts (that is, the need to get multiple things done at once), they felt more time constrained and stressed compared with thinking about their busyness as positive excitement. In other words, choosing to reframe your busyness as a positive state (what the authors call "anxiety reappraisal") can reduce feelings of stress.

To answer Dr. Seuss's question "How did it get so late so soon?": it didn't. You just think it did. And you need to frequently remind yourself of that fact.

III

EVENING

5:00 PM—Exercise

This. Do this.

Do whatever you enjoy. Flipping tractor tires, walking with friends, spinning in the dark to loud music, skipping rope, running stairs, Jazzercise, parkour, Ultimate Frisbee, wall climbing, CrossFit (but I don't need to hear about it), cycling, wrestling . . . Just. Do. Something!

Yes, evidence tells us that particular exercises (such as high-intensity intervals and resistance training) provide more physiological benefits in a more efficient manner than other activities. And some exercise fads seem to be mostly science-free distractions taking up the space of real exercise (I'm thinking of you, hot yoga). But we get the greatest benefit from just being active. So be active. The benefits are so well known and so well established that there is very little to add here. Almost always, exercise is the right decision.

But is after work the best time to exercise? Fitness pundits have much to say about optimizing the timing of workouts, but in reality, the evidence is all over the map on this issue. Some studies suggest that working out in the morning may help with weight loss. There is evidence that afternoon workouts might be best for controlling blood glucose. And evening workouts, at least according to a 2019 study involving mice on a treadmill, are the most productive and result in the best performance. Yet much of this research is less than compelling. Often the data comes from small studies or short-term experiments that find only small differences. And some of the evidence comes from animal studies. This should always lead to a cautious interpretation of the data. Exercise isn't a rat race.

Approximately 80 percent of the U.S. population doesn't get the recommended amount of exercise (which is a modest 150 minutes of moderate aerobic exercise per week, plus muscle-strengthening activities). In

Canada, the numbers are similarly discouraging. And for children, the numbers are even worse, with only 7 percent getting the recommended amount of moderate to vigorous exercise. And we move much less than we think we do. A 2018 study from Canada found that people report that they are physically active about 50 minutes a day but, when objectively measured, it turns out they are active for only 23 minutes. (This is a good reminder to move more than you think you need to move!) Given this grim reality, the overriding priority should be to simply get people physically active anytime—morning, noon, or night. The best time of day to exercise is when you want to exercise.

I usually exercise quite late in the day, often starting after 8 PM. I find this helps me wind down and creates a boundary between work and sleep. (An accumulating body of evidence has found that, contrary to conventional wisdom, evening workouts are not associated with poor sleep.) Don't let pedantic exercise advice—about timing, hydration, breathing, clothing, shoes, supplements, or the number and kinds of repetitions of an activity we should be doing—confuse the simple central message: just move!

5:45 PM—Hang with kids

I am pretty certain my parents loved me. My mom for sure. She was a wickedly funny, quirky bookworm who would have rather argued, martini in hand, about some trivial historical fact involving the British royal family than do our laundry or make our school lunches. Still, I never doubted her devotion. I'm less sure about my father. He was an almost entirely absent—both literally and emotionally—mad-scientist type. I sometimes felt like he viewed my brothers and me more as cheap labor than as his offspring. I spent nearly every summer working on one of his experiments—even from grade one.

I don't believe anyone looking from the outside would have considered awarding my folks a parenting medal. There was very little "parenting" happening, if one defines that as requiring at least some degree of oversight. Here are just a few of the things I did without a word from either parent: organized sword-fighting tournaments that used homemade razor-sharp metal weapons; developed and deployed rocket-fueled mini cruise missiles; dug underground forts that included long, highly collapsible tunnels; and went, from the age of six, on daylong solo adventure hikes around my town, exploring rocky beaches, construction sites, and abandoned houses. There was no overt monitoring of—or strong interest in—my grades, homework, or extracurricular activities. There were no rides to or from friends' houses (I'd walk or not go) or late-night phone check-ins. I'd simply be gone until I got home.

I felt loved and mostly secure. And my memories are of a happy, bustling home.

According to a 2017 survey of 2,000 parents with school-age children, parents feel an average of 23 pangs of guilt about being a substandard parent each and every week. The top reasons for this guilt include "not being home enough," "not playing enough with my children," and "working too much." In a similar survey, conducted in 2018, 57 percent of parents say they struggle to find quality time with their kids and 44 percent say their kids have made them feel guilty about not spending enough time with them. Not unexpectedly, given the pressures and less than equitable distribution of parenting duties, research has also found that working mothers experience far more feelings of guilt than working fathers. And a study from 2017 found that approximately 12 percent of parents experienced full-on "parental burnout"—much of which is tied to a perception of not doing enough.

Parents have likely always felt guilty about something (for most of human existence, simply keeping your kids alive was probably the dominant guilt-driving concern), but now the pressure to be a superparent seems more intense than ever. A mother interviewed for a 2018 study on exhausted parents said, "We are driven by a kind of archetype of a perfect behavior which leads us to set goals that are hard to achieve." Many factors are contributing to this trend, including media portrayals that create unrealistic expectations. A 2015 study, for example, revealed that "exposure to celebrity mom discourse" was associated with a tendency to be a competitive parent and to adopt what the author called an "intensive mothering ideology."

Of course, the rise in social media use has also played a big part. There is growing concern that the sharing of idyllic pictures and stories on platforms like Facebook and Instagram has fostered unrealistic beliefs about what it means to be a parent. A 2017 study involving over 700 mothers explored the influence of social media on parents' attitudes and concerns. It found that, in the U.S., social comparison was a common outcome of social media use and it was associated with a number of negative outcomes, including increased feelings of "role overload" (that is, that there is too much to get done) and lower levels of perceived parental competence.

The message in popular culture is pretty uniform. A good parent is one who spends a lot of time with their children, which might—or even *should*—require sacrificing personal career aspirations. Actress Katherine Heigl reportedly left the popular show *Grey's Anatomy* to spend more time with her children. Musicians Lily Allen and Lauryn Hill took breaks from their successful music careers for similar reasons. Matt LeBlanc quit the international hit *Top Gear* for family time reasons. And comedian and actor Rick Moranis left the entertainment business so he could focus more on his family. I'm not judging these decisions or questioning their sincerity. But the message to the parents of the world: when it comes to spending time with your kids, the expectation is more, more, and more.

As well, many parents may feel that the world is more dangerous than it used to be (even while, as we have already seen, it is in fact far safer than in the past), and therefore more supervision is required.

The core questions, then, are: Do we really need to feel so guilty about how much time we are spending with our kids? And does more "parent time" really matter as much as we think?

First, let's look at how much time we actually spend with our kids. Despite the perception among parents that they aren't spending enough time with their children, parental time with kids has actually increased over the past few decades. In 1965 men spent just 16 minutes a day doing childcare activities. In 2012 that number had increased to 59 minutes. Women still do more childcare work (no surprise), and their childcare time has also increased. According to the Pew Research Center, in 1965, American women spent, on average, 10 hours a week on childcare; in 2016 that number was 14 hours.

These numbers tell two conflicting stories. Yes, for those of us with kids, the increase in hours spent on childcare may contribute to our feelings of an overall time crunch. But they also tell us that we *are* spending more time with our kids, despite a pervasive perception that we aren't.

A 2018 study led by Professor Melissa Milkie of the University of Toronto, and involving data from over 2,000 Canadians, explored how perceptions of having too little time with children may affect parents' mental and physical health. Consistent with other studies, the researchers found that almost half of the parents report that "they feel that they spend too little time with children." But the researchers also found that even though much of the angst about not spending time with children is founded on social expectations—the authors note throughout their paper that parent/child time has increased over the years—this feeling of a "time deficit," as the researchers called it, acts as a genuine stressor on parents, leading to anger, distress, and poor sleep. "Feeling [that they spend] 'enough' time with children is quite important to the well-being of employed parents," the authors conclude. This is true even if the definition of "enough" is set by cultural pressures that may not, as we will see, be tied to the evidence. As the authors note: "It is unclear exactly what employed parents think is remiss, time wise. Ironically, many parents perceive time deficits, and this affects them, even though relative to earlier generations, they spend plenty of time with offspring."

But just what is "enough" time? I have no doubt that part of the motivation for spending time with kids is based on the simple fact that many parents enjoy hanging with their children. I have four children. When I'm away, which is often, I miss them terribly. But the guilt and stress that parents feel are tied to the idea that they are *supposed* to be spending time not for their personal enjoyment but for the welfare of their kids.

This is where the research gets interesting. Despite the deeply ingrained and guilt-inducing conventional wisdom, there is, in fact, little evidence to support the idea that more parent time will necessarily result in healthier, happier, and more thriving children. A 2015 study entitled "Does the Amount of Time Mothers Spend with Children or Adolescents Matter?" followed thousands of kids and found that, in general, the answer was no: the amount of time did not matter to a child's current well-being or well-being measured five years later. This study concluded that "in childhood and adolescence, the amount of maternal time did not matter for offspring behaviors, emotions, or academics." Another study, examining the role of fathers and published in 2016, arrived at a similar result. Those researchers found that when it comes to influencing later child behavior, the amount of time fathers spent with their kids was less important than an emotional connection and a father's positive attitude toward parenting.

Other studies have found that children of working mothers do just as well or perhaps even better than children of stay-at-home mothers. A 2018 study from Harvard Business School that involved data from over 100,000 adults from 29 countries found that adult daughters of employed mothers are more likely to be employed, hold supervisory responsibility, and earn higher incomes than their peers whose mothers were not employed. "The takeaway across decades of research," write the authors, "is that young children of employed mothers tend to be higher achieving and have fewer behavioral problems than young children whose mothers are not employed. . . . These findings add to a growing body of research providing a counterpoint to persistent beliefs and rhetoric that employed women are negatively affecting their families and society."

I want to be crystal clear about the point I'm making here. Of course, parent time matters. These studies do not give us licence to neglect our kids. Safe homes and supportive parents are clearly important. Parental time can benefit a child's development. As Professor Paula Fomby, a researcher on family behavior from the University of Michigan, told me, "The literature is pretty persuasive that parents' time spent with children in specific learning activities like reading is positively associated with children's cognitive achievement." (In fact, reading just one picture book to your child every day exposes them to an estimated 78,000 words a year.) And Professor Milkie told me that time spent with teenagers may be "protective of risky/externalizing behavior."

Despite the messaging that permeates our culture, you don't need to worry about spending as much time as humanly possible with your kids. Granted, this is a topic that is difficult to study well because of all the complex variables involved (socioeconomics, parenting styles, etc.) and the fact that many of the results are correlational. Still, the key point remains: there is very little evidence that more parenting time is always the right answer. So relax.

We also shouldn't allow weird expectations to skew our parenting goals. A 2016 study examined time-use data from 11 Western countries between 1965 and 2012 and found, in general, an increase in parenting time. It also found an association between this increase and parents' education. The authors speculate that one reason "intensive parenting" might be more common among the educated is that it's a status symbol: "their intensive parenting practices confirm their privileged social status by differentiating them from parents in lower social classes." Although more-affluent parents may be engaged in the most intensive parenting, a 2018 study found that all parents, regardless of socioeconomic background, feel that they should aspire to more intensive parenting. The weight of this unrealistic cultural norm may be especially stressful for parents who may already be under financial and time pressures.

We have created a paradox-filled parent-guilt loop. Parents feel they aren't spending enough time with their kids, even though their time with kids has actually increased over the past few decades. This sense of needing to spend more time with kids is likely driven, in part, by a socially constructed idea that that is what good parents do—an idea that may be intensified by parents using social media to post about all the magical time they spend with their kids, thus creating more pressure and/or guilt that, in turn, results in parents spending more time with kids and further increasing parental expectations. All this despite a lack of evidence that more time with kids is always better.

Perhaps we can rethink how we conceive of "quality time." Quality moments—that is, focused and meaningful engagement—don't only involve scheduled face-to-face kid-centric exchanges. Quality time can be found in the everyday activities that families are already doing—running errands, doing chores, and especially, as we will see later, mealtimes. "Quality time" does not necessitate the manufacturing of special moments. Simply living life will often do the trick.

If people want to spend more time with their kids, that's terrific. I want to spend more time with my kids. (Not too much, however. I've got stuff to do!) But in this book I'm exploring the forces that shape our decisions and the evidence behind those decisions. In this regard, we can dial back guilt as a primary motivator for our decisions around hanging with the kids.

5:45 PM—Check phone, again

Don't. If you are trying to have some quality time (there's that phrase again!) with your kids—or, for that matter, with anyone you care about—consider putting your phone away. I don't mean put it on the table. I mean really put it away. We've all heard this for years—but few of us do it. I try. I really do. But I almost always fail. When my phone is in sight, I can feel the pull of my email, social media, and text messages. Almost unconsciously, I'll sneak a glance at the screen. While I was writing this very paragraph, my youngest son came into my home office to tell me about his first high school report card. It should have been a great little moment. But what did I do? I made the decision to look at my damn phone. I did this despite having just spent days and days reviewing the science behind the damaging effect of phones on interpersonal relationships.

Of course, I'm not alone. Ninety-five percent of people admit to taking their phone out in social situations. (The other 5 percent must be sneaking a look in the bathroom.) A shocking one in 10 admit to checking their smartphone during sex! There are many reasons we can't stop doing this. Research suggests, for example, that people who feel most anxious without their phone—a condition called nomophobia—perceive their phone as an extended self. To some degree we all feel this way. Smartphones are both our personal link to the outside world and a repository of the memories that help define who we are. When our phone isn't nearby, we feel something significant is missing.

But despite our understandable attachment to these devices, a growing body of science suggests that smartphones can have a range of adverse consequences for the quality of face-to-face interactions (particularly for that phone-during-sex cohort, I'm guessing).

By allowing us to always be distracted and, in a sense, to be virtually "someplace else," smartphone technology may be reshaping basic human interactions. Consider a study published in 2019 that explored how the presence of phones reduces the tendency to spontaneously smile at strangers. In the experiment, one group of phone-carrying participants entered a room containing strangers and another group entered a similar room without their phones. (The participants were unaware of the purpose of the study.) The participants in the phoneless room exhibited more smiles, including those of the genuine variety. (A genuine smile is called a Duchenne smile—named after the 19th-century French physician and smile researcher Guillaume Duchenne—and involves the use of both voluntary and involuntary facial muscles.) The study's authors speculate that this finding suggests that smartphone technology could be "altering the fabric of social life" by facilitating "our capacity for 'absent presence,' by enabling us to mentally retreat from our immediate environment."

Since this is the time of day that many of us are interacting with our kids—a time when you probably aren't planning to mentally retreat from the immediate environment—it is worth noting that phones can also alter parent-child relationships. The data is grim, but not unexpected. The title of a 2018 study says it all: "Smartphones Distract Parents from Cultivating Feelings of Connection When Spending Time with Their Children."

The act of "phubbing"—checking your phone in the middle of a conversation—is also detrimental (no surprise) to public interactions. Studies have found that as the amount of phubbing increases, the quality of the communication decreases and people feel more socially excluded. Phubbing is particularly hard on romantic partners, as it negatively affects a partner's perception of the quality of the relationship. A 2018 study from India determined that phubbing is so widespread and potentially harmful that drastic action is required, including subjecting the youth of India to special guidance from government clinics to "control this habit in order to promote better physical, mental, and social health."

As is often the case with research on complex human behaviors, we need to be careful not to overinterpret the conclusions. Although a body of research suggesting that phones harm interpersonal interactions is emerging, it isn't easy to create research methods that provide definitive answers on this subject, and there have been mixed results with some of the research. Several well-publicized studies have shown that the mere presence of a phone can reduce the quality of human interactions. A study published in 2016 by a team from Virginia Tech University "found that conversations in the absence of mobile communication technologies were rated as significantly superior compared with those in the presence of a mobile device." Another study, published in 2018, randomly assigned over 300 participants to have their phones either out or not during a social event such as a meal. The phone-out group felt more distracted and, as a result, didn't enjoy spending time with their friends as much as the phone-absent group did. The study concluded: "Despite their ability to connect us to others across the globe, phones may undermine the benefits we derive from interacting with those across the table."

A 2017 study, however, used a similar phone-present/phone-absent experiment but was unable to replicate the "mere presence" result. What they *did* find was that participants' recollections of whether a cellphone was present during a conversation significantly reduced their level of satisfaction with that conversation. If you think it is rude to have a phone out, your memory of the conversation at the time might be worse—regardless of the actual quality of that conversation.

And just to mix things up, some research suggests that the effect of phones on face-to-face interactions isn't all bad. The presence of a phone might help reduce the anxiety of an awkward social situation by acting, as one study concludes, as a "buffer against the negative experience and effects of social exclusion." People who have a phone with them feel significantly less excluded compared with those who do not have a phone. Researchers hypothesize that the devices have this effect for the exact same reason they can be a distraction: phones serve as a connection to others who are not actually present. They are, in effect, a "digital security blanket."

Smartphones are a relatively recent technology. It will be interesting to see if social norms evolve to further accommodate their use while people are in the middle of face-to-face interactions. If so, it is possible that their presence will be viewed as less harmful than whatever the next new technology is—say, neuro implants that allow brain-to-brain communication (BrainTime?).

Despite the uncertainty surrounding some of the research and the evolving nature of social conventions, I think it is safe to provide a few straightforward recommendations that will help us relax and enjoy our time with friends and family a bit more. First, stop the phubbing and put your phone away (out of sight, not just a few inches away) during face-to-face social interactions. If you are waiting to hear from the Nobel Prize committee, an exception can be made, but most of us can survive without a phone check every few minutes.

Second, put your phone on quiet mode. That buzz, ping, or pop-up notification can be very disruptive during one-on-one interactions. "Wondering who has messaged you can be just as bad, if not worse than, taking your phone out and answering it," Kostadin Kushlev, a professor at Georgetown University and lead author of several of the studies described above, told me. He also suggests strategies to limit phone use when in social situations. "To share particularly notable culinary experiences with my boyfriend (who is a total foodie)," Kushlev said, "I will quickly take a photo of my meal, but save the sending for later. Even though sending a photo takes only a few seconds, our research suggests that it might be precisely this brief, intermittent use of smartphones that can fracture our attention and compromise our social experiences."

Third, consider implementing phone-free times during the day, particularly when you are finished working and trying to wind down, exercise, or hang with friends and family. Or at least don't carry your phone with you. A strategy that I use—with varying degrees of success—is to leave my phone in my home office when I get home from work. If I want to check it, I must go into my office. That small obstacle will force me to question whether I really need to check Tom Brady's updated stats at that exact moment.

Fourth, recognize your triggers. If you have an urge to check, ask yourself if you really need to look at your phone right this instant. What is compelling the urge? Boredom with your friends and family? Stress about work? A desire to catch up on gossip? Are any of these reasons worth retreating from the immediate experience?

6:00 PM—Dinner

This decision is easy. If at all possible, err on the side of eating a home-cooked meal. And, if you have one handy, eat it with your family—or your romantic partner or a buddy or a cat or favorite plant.

It probably comes as no surprise that studies have consistently demonstrated that eating meals at home is associated with better food choices, including the consumption of more fruits and vegetables. The benefits are even more pronounced and wide-ranging when it comes to young people. Children and adolescents who share family meals are less likely to be overweight and more likely to eat a healthy diet. (In contrast, frequently eating meals in front of the TV, for example, is associated with being overweight or obese—likely because, among other things, it promotes mindless eating.) A 2018 study found that even if your family is, as the researchers put it, "dysfunctional," eating a family dinner together is still associated with improved dietary intake.

Family dinners can also promote other healthy eating habits, including eating fewer calories. In a world of expanding portion sizes—one of many trends associated with the current obesity crisis—family dinners are an opportunity to revisit what an appropriate amount of food looks like. Even children as young as two can become accustomed to eating inappropriately large portions. A 2018 study found that, at a family meal, "reducing food portion sizes may recalibrate perceptions of what constitutes a 'normal' amount of food to eat and, in doing so, decrease how much consumers choose to eat." Family dinners can help guard against the market-driven trend of massive portion sizes, giving both kids and adults the skills to recognize an absurdly unnecessary supersize when they see one.

The benefits of a family dinner on the welfare of youth go well beyond simply improved nutrition and portion control. A 2011 systematic review of the evidence concluded that frequent family meals are "inversely associated with disordered eating, alcohol and substance use, violent behavior, and feelings of depression or thoughts of suicide in adolescents." The study also found that they are positively associated with increased youth self-esteem and school success. Family dinners are also linked to improved vocabulary and family bonding and a greater sense of resilience.

What's more, these benefits appear to endure for years. A 2018 study from the University of Montreal followed a cohort of almost 1,500 children from when they were 6 until they were 10. The researchers controlled for a range of variables, including socioeconomics and cognitive abilities, and found that the quality of the family-meal environment at age six predicted higher levels of general fitness and "lower levels of soft drink consumption, physical aggression, oppositional behavior, nonaggressive delinquency, and reactive aggression at age ten."

Family meals are also an opportunity to teach children about food and cooking, skills that will provide benefits for a lifetime. To paraphrase that old proverb: Give a child a meal and you feed him for a day. Teach a child to cook and he can avoid McDonald's forever.

All this data is impressive, and although a lot of the research is correlational in nature, this topic has been explored from a variety of methodological angles, and the results are uniformly and consistently positive.

"We'll never be able to causally prove the merits of family dinners," Dr. Yoni Freedhoff, a professor in family medicine at the University of Ottawa and an expert on weight management, admitted when I asked him about the causation/correlation issue. "But there are a wide breadth of positive associations, from improvements to mood, school, and dietary quality to decreases in teen substance abuse and risk-taking behaviors. Plus a total lack of negative associations," Freedhoff said. "It would seem that eating meals together as a family, around a table, ideally free from distractions, is something worthy of striving for in every home."

Dr. Sara Kirk, a professor of health promotion at Dalhousie University, agrees with Freedhoff. "There is sufficient evidence to support family meals as important for a range of health and psychosocial outcomes," she said. Kirk's research explores ways in which we can, as a society, create healthy environments in order to help prevent chronic diseases. Looking at what the evidence says about family dinners is right up her alley. Like Freedhoff, Kirk noted the challenges around trying to establish causation, but feels the evidence of the value of family dinners is, taken as a whole, pretty conclusive. "Some of the studies from which these findings are derived are in large population samples," she said. "That helps to increase our confidence in the findings."

To be fair, this advice, for some, falls into the category of "easier said than done." Single parents and people who work shifts, for example, may find it a challenge to regularly sit down to a family meal. And this is exactly what a 2017 study from the University of Oxford found: there are demographic differences in how people approach dinnertime. Families with higher-educated members have more frequent family meals. Those that are more affluent spend more time at the table. And people with challenging schedules, most notably single parents, "spend the least amount of time eating with their families." Given these socioeconomic realities, public health experts have been exploring ways, on a population level, to encourage family meals, including providing families with classes on home cooking and having employers offer cash incentives to encourage healthy eating patterns. The results have been mixed. Encouraging meaningful, sustained behavior changes is tough, particularly around something as complex and culturally mediated as family meals.

"The obstacles to family dinner are familiar: not enough time, money, or cooking skills, picky eaters and tension at the table," Lynn Barendsen told me. She is the executive director of Harvard's Family Dinner Project. "To us, the goal is not perfection—the 'good enough' dinner really is good enough. We suggest families start small: set one goal, choose one thing to work on first."

The project is designed to encourage family dinners as an opportunity for family members to connect with each other through food, fun, and conversation about things that matter. "More than 20 years of research tell us why family dinners are important," Barendsen said. "Magic can happen without perfection. It doesn't take long!"

Many of the benefits associated with eating a meal with friends and family, particularly those of the psychosocial variety, likely flow from the communal nature of the activity. This raises the question: Are all the advantages lost if you eat alone? An emerging paradox, relevant to us all, was emphasized in my exchange with Dr. Kirk. "It is interesting," she told me, "that just as the research evidence grows on the importance of family meals, or more broadly of meals shared with others, we are seeing an increasing trend in people eating more meals alone, both in and outside the home. The reasons for this are, of course, complex, and reflect societal changes such as an aging demographic, more people choosing to live alone, increases in rates of separation and divorce, work patterns, and numerous others."

Some research, including a 2018 study from Korea of almost 8,000 people, found that eating alone was associated with poorer health. A 2015 study from Japan associated eating alone with unhealthy dietary habits. The challenge, from a research perspective, is to disentangle the other factors associated with eating alone—particularly loneliness, which we know can have a profound effect on our health. Is it the loneliness or the eating alone that is creating the health problems? It is likely a combination of many factors that leads to the poor health outcomes. (It is worth noting that public health officials are justifiably focusing more attention on the critical issues associated with social isolation, as highlighted by the U.K.'s appointment of a minister of loneliness.)

To be sure, not everyone wants to be involved in a communal meal. One study of French and German young adults found that many of them prefer eating alone. Sharing meals can be stressful, or eating alone can be a time to unwind. If you decide to eat alone, as many of us do, focus on the nutritional and relaxation benefits that can come with eating a healthy home-cooked meal.

Throughout my decades-long career weeding through research relevant to health policy, very few interventions have felt as sensible as the idea of eating meals with people we love. After all, this practice is at the core of the human experience. But am I allowing my interpretation of the science to be swayed by the natural appeal of the premise?

For a number of years, my family has been fortunate in being able to spend one solid month every summer traveling together. We love settling into the pace of whatever city we are in. We don't scramble to cram every moment with sightseeing. And each night we all sit down for a family dinner. Free of the usual obligations of work, friends, and extracurricular activities that crowd our schedules at home, these dinners often last three or four hours. And thanks to oppressive international roaming fees, there is not a cellphone in sight. We argue. We share gossip, fears, insecurities, and aspirations. We laugh until we cry. These dinners are glorious. They are the favorite moments of my entire life. That sounds like an over-the-top pronouncement. But it isn't. If an alien abducted me and said I could keep only one set of memories, it would be these dinners—and, as a bonus, the Patriots' 2017 Super Bowl win. Obviously.

6:15 PM—Wine

Let's start with the conclusion on this one. Wine tasting is mostly nonsense. By "wine tasting" I mean the ability to objectively and reliably discern a quality difference between products. And by "nonsense" I mean people can't. This reality is great news for people who enjoy wine. Simply buy what you like. Ignore the price, bottle, label, history, region, and even color.

The global wine market is approaching $400 billion. Much of the marketing for this massive industry plays on the subtle differences between varieties and price points. Walk into almost any liquor store and a wall of wines, often broken down by country and region, will confront you. There are wine-tasting clubs, wine-tasting magazines, and, of course, wine-tasting experts, many of whom likely went to a wine-tasting school. There are wine-tasting board games. And one company, Vinome, will even test your DNA to determine what kind of wine you are genetically predisposed to enjoy (a scientifically absurd idea and one I've tried in order to fully experience its scientific absurdity-ness). Part of the fun of being a wine drinker—or so I've been told, as I'm an overpriced-beer man—is the search for new and better wines. And the industry plays to this appeal.

But the science on whether we can tell the difference between wines is pretty consistent. We can't. At least not very well.

Probably the most famous study on point was published in 2001 in the journal *Brain and Language*, and it rocked the oenophilia world. The researchers gave tasters white wine that had been dyed red. The tasters couldn't tell the difference. Many described the white wine as if it were a red. They were completely fooled. To make matters worse for the wine world, the tasters were studying wine at the Faculty of Oenology at the University of Bordeaux. Ouch.

Since that experiment, many others have further supported the conclusion that wine tasting is of questionable legitimacy. One study, for example, explored why the results of wine competitions are so inconsistent (which, in itself, is a pretty big hint that things are amiss, tasting-wise). The researchers were curious: "Why is it that a particular wine wins a Gold medal at one competition and fails to win any award at another?" To answer the question, they followed dozens of judges at a major wine competition between 2005 and 2008. Each panel of judges reviewed 30 wines that, unknown to them, included three samples poured from the same bottle. Only 10 percent of the judges were able to replicate their score within a single medal group. The results are far from impressive, especially for wine judges. "Chance has a great deal to do with the awards that wines win," the author of the study, Robert Hodgson, told the *Guardian* in 2013.

You might be thinking, "Okay, people can't tell the difference between a variety of high-end wines, but people know the difference between expensive and cheapo wine." Again, mostly not. One blind taste test experiment in the U.K. had 578 people try a variety of cheap and expensive red and white wines. The researchers asked them to declare whether the wine was expensive or inexpensive. People were correct 53 percent of the time for white wine and 47 percent for red. They would do as well by flipping a coin. Another study, published in 2008, that involved over 6,000 blind taste tests found that there was no relationship between the price and the rating of a wine. In fact, the researchers found that, on average, "people enjoyed the more expensive wine slightly less."

To be fair to the world of wine, a 2009 blind tasting demonstrated that humans are also incapable of telling the difference between fancy pâté and dog food. (Seriously, researchers did this study!) The dog food was rated as poor-tasting pâté.

As you can imagine, wine fans are not keen on the anti-wine research and have tried to poke holes in the experimental approaches. But such quibbles are irrelevant. If there were clear and robust differences between wines, they would be easy to detect using a range of experimental designs. Also, a rich body of literature demonstrates that objective quality often has very little to do with how we assess a range of products. A 2018 study, for example, found that non-taste factors, such as where a wine is from, influence how much we are willing to pay for a bottle. The researchers found that revealing that a wine was from, say, Iowa caused people to value the wine less than if they didn't know where it was from. The context in which you drink wine can also influence your drinking experience, including the location (fancy restaurant vs. kitchen table) and what music is playing in the background. Nor are experts immune to such influences. A 2018 study from the University of Oxford found that "experienced wine tasters were also influenced by background music when it comes to wine evaluation" and that "years of wine tasting experience do not moderate the impact of music on wine evaluation."

Angular, *toasty*, *foxy*, *fleshy*, *opulent*, *steely*, *flamboyant*, *unctuous*, *intellectually satisfying*—the world of wine words often seems ridiculous. But the words work. People's preferences can be manipulated by how a wine is described. In a 2017 study, researchers gave participants the same wine under three different conditions: blind tasting with no information about the wine, tasting with basic information, and tasting with elaborate information. No surprise, the elaborate information resulted in the highest liking and positive emotions toward the wine and a "substantial increase in willingness to pay after tasting." There's money in that complex bouquet of rich, oaky verbiage!

We often purchase, like, and believe things simply because they fit with our personal brand and how we see ourselves. So it shouldn't be a surprise that the design of the wine label can sway our opinion. A 2019 study from the University of British Columbia found that if people like the label, and particularly if they identify with the vibe of the label, they are more likely to enjoy the wine. If you think of yourself as fun and a bit zany, then you will find that a bottle of wine with a fun and zany label tastes great.

But perhaps the biggest influence on our subjective assessment of the quality of something is price. If something costs a lot, we like it better. Several studies have explored this phenomenon as it applies to wine. A 2011 experiment revealed that price can actually be used as a marketing tool, because "disclosing a high price before tasting the wine produces considerably higher ratings" and "influences expectations that in turn shape a consumer's experience." In 2017, researchers at the University of Bonn used MRI scans to look at how price affects the perception and brain activity of wine drinkers. A price would flash on a screen and then a bit of vino would be delivered to the participant via a tube. (The same wine was given to all the subjects.) The higher price made people rate the wine as better tasting. The researchers found that the parts of the brain associated with reward were more activated by seeing higher prices, thus priming expectations. And this sensory expectation had an actual biological effect that made the wine taste better. This has been called the "placebo effect of price," whereby expensive products are perceived as better even if they are identical or nearly identical to cheaper versions. This happens, to a greater or lesser extent, with a vast range of consumer products, such as sound systems, face creams, tea, cheese, cellphones, jeans, and vodka. Even the perception of my most often purchased beverage, my beloved coffee, is heavily influenced by price and context. I couldn't find any academic studies on point, but there are many blind taste tests cited in the popular press. One taste test had Folgers instant coffee ranked second and Dunkin' Donuts third, followed by several more niche exclusive brands, with Starbucks coming up last. Blue Bottle coffee, a mid-price product, won the test. A blind taste test done by the *Today Show* pitted a one-dollar deli coffee against brew from a gourmet coffee-house. The winner, preferred by 67 percent of the participants, was the cheap coffee. And in 2012, *Jimmy Kimmel Live* punked people on the street by fooling them into thinking cheap deli coffee was a cup of Starbucks' more high-priced (seven dollars for a cup!) product. Of course, there are many factors that impact our purchasing decisions for things like coffee and wine—including personal branding—but price clearly has a big influence. For many products, once the price is removed, the illusion of difference falls aside.

One of my favorite examples of how context can alter our perception of quality is a masterful marketing stunt done in the fall of 2018. The discount shoe company Payless opened a fake luxury store, called Palessi, in a former Armani shop and staffed it with models. At the grand opening event, customers thought they were in an exclusive, high-end designer boutique and praised the fine workmanship, the high-quality materials, and the sophisticated look of the shoes. They were willing to pay hundreds of dollars for them—even though they were the same $20 or $30 shoes they could purchase at Payless. Because of the fancy atmosphere, the customers were prepared to pay hundreds more than the actual retail price—as much as $640, which is an 1,800 percent increase!

Of course taste and preference are influenced by subjective factors. We have always been and always will be swayed by the context in which we buy and consume things. Indeed, for many goods, like wine, part of the pleasure derives from the history and culture. The backstory matters. It can be difficult to escape the pull of branding and the illusion of difference, but reminding ourselves that external biases can so easily sway us can afford a kind of freedom, liberating us from the marketing noise and, perhaps, saving us lots of money.

Confession time. On the weekends my wife and I often enjoy an ice-cold tumbler of vodka as we kick back for some TV bingeing. We aren't big drinkers, so it can take us many months to get through a single bottle. But even though I know that blind taste tests show, over and over again, that people can't tell the difference between cheap and expensive versions of a product, and even though I know about the powerful sway of appearances, I always buy the same expensive Polish brand. Why? The bottle is so fancy!

Before we leave the topic of alcohol, you may be wondering whether you should even be drinking it. The general public has been subjected to a considerable amount of flip-flopping on whether a moderate amount of alcohol—one or two drinks a day—is safe or even healthy. Unfortunately, much of the relevant research that has informed the sensationalistic headlines about both benefit and risk—"There's No Safe Level of Alcohol"—is merely observational in nature. As a result, we can't say definitively what should be considered, from a health perspective, an ideal and/or safe amount. What we do know: heavy drinking is really bad for you. Alcohol is one of the deadliest substances humans consume, resulting in three million deaths worldwide every year, including those associated with alcohol-induced car accidents, violence, and disease. Given this, any recommendation about alcohol needs to be tempered by the knowledge that excessive consumption causes significant harm. If you do drink, it should be in moderation, which the current body of evidence and most public health agencies—including the U.S. Centers for Disease Control and Prevention—suggest is one drink a day for women and two for men. There are recent studies, including the 2018 study published in the journal *The Lancet* that generated the above "no safe level" headline, that hint that these recommendations are too high. Until we have better studies that can tease out causation—such as a large, well-done clinical trial—I think it's safe to assume that, if done responsibly, moderate drinking is probably fine. But we are a long way from declaring wine, or any alcoholic beverage, a health food.

7:00 PM—Dishes

Another easy decision: do the damn dishes. Your romantic partner will appreciate it greatly. This advice is primarily aimed at men in a heterosexual relationship because research tells us we are particularly bad at sharing chores—though we are getting a bit better. But I'm sure this is sound advice for any couple. In fact, recent studies suggest that a more egalitarian distribution of dish duties leads to more frequent and more satisfying sex. A 2016 study from the University of Alberta found that when "male partners reported making a fair contribution to housework, the couple experienced more frequent sexual encounters, and each partner reported higher sexual satisfaction one year later." And if you want to get the biggest bang, so to speak, for your buck, it makes sense to focus on doing the dishes, which is the most reviled of all chores. A 2018 study from the University of Utah found that "the division of dishwashing, among all tasks, is most consequential to relationship quality, especially for women." In 2018, the U.S.'s Council on Contemporary Families, an academic group exploring the latest research on point, summarized the dishes data thus: "Women who found themselves doing the lion's share of dishwashing reported significantly more relationship discord, lower relationship satisfaction, and less sexual satisfaction than women who split the dishes with their partner."

Of course, these studies are really about a fair distribution of chores. And a decision to do the right thing—that is, sharing grimy tasks—shouldn't be driven by the hope of carnal pleasures. Still, improving your sex life seems a solid justification for picking up that sponge.

7:30 PM—Toilet seat up or down?

I'm sure the great toilet seat debate has raged ever since John Harington invented the flushing toilet in 1596. The central point of contention: Should men put the seat down after peeing? As more and more gender neutral facilities are popping up in public places, including offices, restaurants, and schools, the need to resolve this debate is likely to intensify. And for many, it's already pretty intense. In 2014, for instance, a North Dakota man was arrested for assaulting his sister after she complained about him leaving the seat up. That same year, a Pennsylvania woman violently murdered her husband—she stabbed him 47 times—after he had "forgotten to put the toilet seat back down after using the bathroom." (Okay, that last one is from a fake news website, but the fact that it feels like it could be true speaks to how high the toilet seat stakes have become.)

Before I commence this analysis, let me clarify: I'm not referring to men peeing on the toilet seat. I get that this is a bad thing. In my experience, any mention of men and toilet seats seems to provoke rage about urine splash. I get it—it's unpleasant. Incidentally, women pee on toilet seats too. Pee on the seat is a truly universal and, apparently, gender-less nuisance.

Then there is the proposition that all men should sit for all forms of toilet activity. Indeed, in some countries there has been an official push to get men to sit. In 2012, for example, the Taiwanese environmental protection minister suggested that men should sit in order to help promote a cleaner environment. A political party in Sweden tried to make it illegal for men to stand while urinating in public washrooms—again, arguing that sitting is cleaner and more hygienic. Given this kind of activity, it is no surprise that sitting is becoming more common. A 2018 survey from Japan found that 33 percent of men prefer to sit and that 44 percent sit to pee while at home. But perhaps an even more telling bellwether: there are, um, leaked reports that Ryan Gosling sits to pee. If a manly, trendsetting heartthrob like Gosling has opted to sit, then surely the rest of us will follow.

In Germany, a court adopted a different posture on pee etiquette in a 2015 legal case that ruled that men have a right—a *right*, dammit!—to stand while urinating. The case involved a landlord alleging that a tenant's urine spray did damage to an apartment floor and insisting that the tenant sit. In finding in favor of the defendant, Judge Stephan Hank seemed to lament the potential passing of an era. "Despite the increasing domestication of men in this area, urinating while standing up is indeed still common practice," he wrote in the ruling. And in what feels like a superfluous aside drawn from personal experience, Judge Hank also noted thus: "Someone who still practises this previously dominant custom is regularly confronted with significant disputes, particularly with female cohabitants."

Despite these intriguing sociopolitical developments, here I will keep the focus on the purest form of the toilet seat debate: Is there an evidence-based argument that supports the contention that the seat should always be put down?

I did a very unscientific social media survey on whether the toilet seat debate is really about etiquette (putting it down is the polite thing to do) or efficiency (it makes sense, in the aggregate, from an evidence-based, time-saving perspective). This request for the public's perspective generated 1,300 votes and more than 100 comments in just one day. Most respondents (62 percent) said this was about etiquette. That surprised me, because if it is a matter of etiquette, then this is not a question that can be resolved by evidence. Rather, asking men to put the seat down is akin to saying men should hold the door open for women or hold their chairs when they sit. A wee bit old-timey, methinks.

As argued by writer Jonathan Wells, a more progressive approach would be the one-touch rule. "Everyone agrees to touch the seat once during their lavatory trip," Wells wrote in the *Telegraph*. "Men put it up at the start, women put it down. What could represent a more equal compromise, I humbly ask?" Given that no one really wants to touch the toilet seat, if this is strictly about equality, then sharing the up/down responsibility seems the fairest approach. It is courteous to everyone involved.

But one compelling counterargument to the Wells proposal—one that nudges us closer to an argument based on evidence—is the "you are touching it already" theory. A 2015 *Bustle* article that passionately defended the men-put-the-seat-down proposition noted that since men need to touch the seat anyway, why make women touch it too? By asking men to have the exclusive seat-moving responsibilities, we are reducing the number of humans who need to touch a toilet seat. Since everyone ought to wash their hands post pee, this is mostly about convenience and time and, I guess, the yuck factor. Which brings us to the efficiency question.

At some level, if it isn't about etiquette, it is about personal cost. Moving the seat up or down involves time and effort, which makes it a quantifiable issue that can be analyzed systematically. And, believe it or not, Dr. Jay Pil Choi, the distinguished professor of economics at Michigan State University, has done this analysis. In a 2002 paper filled with mathematical formulas and complex graphs, Choi argues that, in most situations, efficiency favors what he calls the "selfish position," which is leaving the seat in whatever position it is in when you are done (basically, the one-touch rule). This approach reduces the total number of toilet seat moves. Unless there is a large asymmetry in the ratio of users, Choi argues that the math does not support the seat-down convention. A game theory analysis done in 2007 by Hammad Siddiqi, a senior lecturer in economics, came to the same conclusion. His analysis found that "the social norm of leaving the toilet seat down is inefficient." Again, this is considering only the time cost and inconvenience of moving the seat. But Siddiqi also did an analysis where he tried to consider the cost of having an angry romantic partner and determined that when this is folded into the equation things get, no surprise, much foggier. Basically, he determined that it probably isn't worth the fight, even if the social norm is not supported by the math.

Can we put an end to this bathroom battle? Not really. From a purely evidence-based perspective, the rationales for always putting the seat down are pretty weak. But this evidence-is-equivocal conclusion still makes an important point. Few topics covered in this book stirred as much anger among my friends and colleagues as this one. Not the bit about kids walking to school. Not ranting. Not raw milk. Nothing compared to the rage I heard about toilet seat behavior.

I'm almost certain that my underwhelming conclusion won't satisfy anyone. The toilet seat debate feels like a proxy argument for other, more substantial gender issues. So the science and logic probably don't matter. But whether you are talking about etiquette or efficiency, I can't find a slam dunk argument that would justify the intense passion this topic invokes.

Nevertheless, here is the most reasonable and evidence-based approach: if a toilet has a lid, then the best thing to do is close the lid before you flush. This will limit the toilet plume and, yes, require the lowering of the seat. As a result, we will all, both standers and sitters, need to lift the lid to do our business. For the standers, this leaves us in the same place, from an efficiency point of view, as being asked to lower the seat. However, for the sitters, this adds an additional move (always having to lift the lid). So, in some respects, the sitters lose with this approach because now they will always have to move something and, arguably, it is slightly more difficult to lift a lid than to lower a seat (gravity and all). Still, despite this inequity, if the comments I received in response to my survey are any indication, most people feel happy with this resolution.

But will this up/down détente finally put a lid on the great toilet seat debate? Seems unlikely.

7:50 PM—10,000 steps?

This is the time when most of us have finished the physically active portion of our day. Bring on the sofa! But you may be wondering: Did you do enough? Did you move as much as you should have moved? Specifically, did you take 10,000 steps?

There is nothing magical about 10,000 steps. This widely accepted and oft-repeated goal is completely arbitrary and, in fact, seems to have originated in a 1960s marketing campaign from Japan aimed at selling one of the world's first step counters. Of course, more steps per day are generally considered a very good thing. But there is no science associated with the 10,000 norm. It's just a catchy and easily remembered number.

Yet even though this step goal isn't backed up by deeply researched public health data, it has taken on an incredible amount of cultural momentum. Entire industries, like activity tracker technology, are built around the idea of getting 10,000 steps. I love anything that gets people active. But it isn't clear that activity monitors fulfill this promise. Here's the evidence-informed reality on activity trackers:

- They are often inaccurate.
- They usually don't do a great job accounting for intensity.

- They don't help people lose weight.
- The evidence on their sustained use and motivation to exercise is mixed (e.g., one 2018 study found that 57 percent thought their activity tracker increased their exercise despite the fact that it had actually declined).
- For some populations, like adolescents, they can lead to less physical activity.
- They can make exercise less enjoyable.

This last point is key, and is certainly my experience. I find that worrying about metrics—speed, calorie count, distance, and so on—robs me of the joy of the activity. Ideally, physical activity should be something you like and can sustain over a lifetime. While some people find activity trackers motivating, there is at least some evidence that they may make exercise more of a grind. And this could erode the other benefits associated with exercise, such as stress reduction. As noted in a 2016 study from Duke University, "By drawing attention to output, measurement can make enjoyable activities feel more like work, which reduces their enjoyment. As a result, measurement can decrease continued engagement in the activity and subjective well-being."

My recommendation: less measuring, more enjoying.

8:00 PM—Binge-watch TV

One of my favorite coffee shops in Toronto doesn't have its own bathroom. You need a code to use one shared by a connected office space. This particular bathroom has a single small stall, and about half the time I visit (well, perhaps it just *feels* like half the time) there is someone sitting on the toilet streaming a TV show.

Bingeing has been defined in a variety of ways, such as watching an entire season of a show within one week or watching numerous episodes of a show at a single sitting. But I think all can agree—especially given the behavior of Mr. Toilet above—that it has become a ubiquitous phenomenon. According to a study done by Deloitte, 70 percent of Americans admit to binge-watching TV. For Generation Z, a cohort defined as people between the ages of 14 and 20, that number rises to 90 percent. Other research has put the number of bingers even higher. A 2017 study from the University of Leuven found that over 80 percent of the general population classify themselves as a binge-viewer. And many watch at a pretty intense pace. When Netflix released season 2 of the popular show *Stranger Things*, almost 400,000 people watched the entire season, about eight hours of TV, in a single day. There is even a new category of streaming superfans, the "binge racers": people who view bingeing—and finishing a series as quickly as possible—as a kind of competitive sport. "There's a unique satisfaction that comes from being the first to finish a story," Netflix's vice president of original series, Brian Wright, said. In 2017, Netflix estimated that over 8.4 million of their customers had chosen to binge race at least once.

As Mr. Toilet TV illustrates, TV viewing now happens throughout the day, whenever and wherever you want. I'm pretty sure some of my students may be watching TV *during my class*. But most people binge in the evening. A 2018 industry survey of 1,300 Americans found that nearly half admitted to pulling an all-nighter in the past year in order to binge a TV show. An incredible 85 percent said they stream while in bed (spoiler alert: *do not do this!*), with divorced people being the most likely demographic to engage in this nocturnal viewing. And 73 percent of the public said they sometimes fall asleep while watching streaming content.

Given the growing tendency—and I would suggest near-total social embrace—of bingeing, it is no surprise that researchers are exploring the phenomenon. This has included speculation about why we have the urge to binge. A small 2014 study, for example, found that people form close ties with TV characters and that this (obviously one-sided) bond is one of the "main factors influencing bingeing behaviors." In a 2017 study, 59 percent of respondents said they binge simply because they like to see the whole show at once. Another half admitted that they don't like the suspense of waiting to find out what happens. Interestingly, 30 percent admitted that they binge merely to fill time—a curious response given how busy we all claim to be.

But other, less obvious, forces may be making us binge. Some researchers have suggested that our brains are wired to watch. We simply can't stop ourselves from starting the next show. Like many other pleasurable activities—eating chocolate chip cookies, for instance—there is always room for just one more. And the streaming services know this. They have made moving on to the next show the default behavior. There is a nudge to continue. You don't even need to reach for the remote. The next show loads automatically. And when you add the tendency to have shows end on a cliff-hanger—a narrative tool that creates psychological tension begging for a resolution—the desire to continue watching can be tough to fight.

Before I outline why all this bingeing may be problematic, I must confess that I totally get it. Zero judgment from me. This is, as has often been said, the golden age of television. Still, evidence is emerging that bingeing has health consequences. For example, a 2017 study of 423 adults found that a higher level of "binge viewing frequency was associated with a poorer sleep quality, increased fatigue, and more symptoms of insomnia." The researchers speculate that this is probably due—in addition to the simple fact that bingers stay up later—to the increased presleep cognitive arousal caused by the immersive experience. The intense engagement with the characters and story may mean it is tougher to relax and transition to sleep mode. As I have already noted in this book, most of us aren't getting enough sleep, so any trend that further deprives us of sleep time cannot be viewed as entirely positive.

Binge-watching has also been associated with a host of poor lifestyle habits, including unhealthy snacking, not eating enough fruits and vegetables, consuming more fast foods, and (zero surprise here) too much sedentary behavior. A 2017 study from Brigham Young University randomly sampled 500 young adults and found that, in general, bingeing was linked to many of these kinds of less than ideal behaviors. Among the binge-watchers, 85 percent ate fruit and vegetables less than once a day, 88 percent ate out once a week or more, and less than 50 percent met the physical activity recommendations. And a 2017 survey of U.S. undergraduate students suggests that bingeing may be a major cause of schoolwork neglect: 63 percent of the respondents identified binge-watching as an academic obstacle.

There is also concern that bingeing could increase the risk of getting a blood clot. A study from the University of Minnesota published in 2018 followed over 15,000 people for 24 years and found that excessive TV watching was "independently associated with increased risk of VTE"—that's venous thromboembolism, also known as deep vein thrombosis, which is a blood clot deep in the tissue. The researchers found that even those who were getting the recommended amount of daily physical activity faced this risk. Another study, published in 2017, suggested that TV viewing was associated with (again, emphasis on *associated with*, which means only a correlation was found) an increased risk for a range of inflammatory conditions, including Alzheimer's disease, type 2 diabetes, and kidney disease.

Bingeing has likewise been linked to mental health concerns. It has, for example, been connected with feelings of loneliness and depression. A 2015 study of more than 400 adults found that those who binged regularly "reported feeling more depressed and anxious than those who spent less time watching TV." And more than half of the 2,000 respondents in a 2018 U.K. study "admitted to having experienced mental health issues brought on by the end of a TV series."

The mental health consequences may relate to how we watch. A 2018 study from the U.K. found that most TV viewing (75.8 percent) was done alone. Even cohabitating couples usually watched alone, which sounds genuinely sad. A 2016 study of Dutch binge-watchers also found it to be a solitary activity. The flexibility of streaming probably contributes to solo bingeing, as you don't have to coordinate with other people to watch a certain show at a specific time. As such, streaming platforms might be causing more isolation. It is even less of a communal activity than old-school TV watching.

It is also worth asking whether bingeing is the best way to watch a show from the perspectives of entertainment, comprehension, and long-term appreciation. Most people say they enjoy binge-watching—so that's obviously one reason people do it. And, as noted, some people also enjoy the satisfaction of watching an entire season as fast as humanly possible. Getting through a show quickly—and before most of the world—feels like an accomplishment, one that will also protect you from, and allow you to become the source of, spoilers. But some research suggests bingeing might actually reduce the quality of the viewing experience. A study done by the University of Melbourne in 2017 involved randomly assigning participants to watch a TV show once a week, daily, or bingeing in one sitting. What it found was that people got the most enjoyment out of watching a show daily and the least enjoyment out of bingeing. The binge group also had the worst recall and comprehension (that is, they performed more poorly on questions designed to test their memory and understanding of the show). If you really want to get the most out of a show, bingeing might not be your best option.

I have several big caveats before I get to my bingeing recommendations. If you've gotten this far into the book you can probably guess what they are.

First, the research regarding harm is far from conclusive. I would categorize most of it as exploratory. This area of study is relatively new. Most of the data we have now, particularly in the context of adverse outcomes, is correlational. Although a consistent body of evidence is starting to surface, we shouldn't overplay the stated risks. For example, we don't know whether lonely and depressed people are more likely to gravitate to bingeing or whether that bingeing facilitates loneliness and depression. It is likely a combination of both. One study, published in 2018, found that people who were primarily motivated to watch for entertainment purposes—as opposed to, for example, killing time—were more likely to have a low level of binge-watching. In other words, those who were already watching for what might be considered the most constructive reason—enjoyment of the program—were already the ones doing it in the most responsible manner. This demonstrates how the causal connection between bingeing behaviors and wellness outcomes can be complex.

Second, we need to remember that humans have a tendency to see harm in almost every new medium. Socrates was not a fan of writing because it "will create forgetfulness in the learners' souls; because they will not use their memories; they will trust to the external written characters and not remember of themselves." There was concern that the printing press would degrade our ability to think clearly. As noted by historian Roy Porter, there were many "post-Gutenberg misgivings about the perils of print." It might, as scientist Conrad Gessner speculated in 1565, lead to information overload that the public would find "confusing and harmful." In the 18th and 19th centuries there was concern that women wouldn't be able to tell the difference between reality and the vivid worlds portrayed in novels.

Similar misgivings were raised about radio, comic books, movies, good old-fashioned TV, video games, and the internet. All of these technologies transformed our society, but the harms were almost always overplayed or, at least, oversimplified. We should remember this whenever we hear pontificating about how a new media technology is destroying our mental and physical well-being. Some degree of concern is justified, but we should all moderate our worry with a healthy dose of critical thinking.

Would I judge Mr. Toilet TV differently if he were in that stall reading Tolstoy? Probably. How about if he "binged" Wagner's epic 15-hour Ring Cycle opera over four nights at the Metropolitan Opera House?

Binge-watching has, in a remarkably short time, become a cultural norm. It is how many of us now watch TV. So it is entirely fair to consider the social, physical, and mental health implications. Poor sleep and too much sedentary behavior are already well-known public health issues. Both common sense and the emerging evidence suggest that bingeing may be making things worse—or, at least, making it more difficult to make things better. The psychological implications of the trend are, at this time, more speculative. But, again, the evidence is suggestive.

Dr. Stan Kutcher was a professor of psychiatry at Dalhousie University and is a renowned expert in adolescent mental health. He told me, "Although the science is just beginning to untangle the complex relationship between excessive watching of TV and health and mental health outcomes, excessive use may lead directly to some negative outcomes, such as challenges with emotional regulation"—that is, how we respond to an emotional situation—"or indirectly through replacement of other activities that promote health and mental health, such as face time with friends and families, exercise, etc."

Dr. Kutcher makes a key point about how bingeing displaces other activities. So many of the suggested harms are tied to this problem. As such, when it comes to bingeing, the recommended way forward is fairly obvious: moderation. While we should probably relax about the more dire warnings, we do need to find a viewing strategy that is healthy for us.

"How much is too much may differ from person to person," Kutcher said. "A key question that should be asked is, How is my use of

TV—and other screen time—promoting or detracting from my physical and mental health? An honest answer to that question should help you decide if what you are doing should change."

But it is tough to make a rational decision about the need for more sleep or exercise or study time or family time when you are dying to know what happens next on *Game of Thrones*. What we need are strategies that will help us strike a healthy balance.

A number of experts, including the American Academy of Sleep Medicine in a 2017 statement on binge-watching, have put forward sensible suggestions for how to minimize excessive TV watching, including: decide beforehand how much you are going to watch and do your best to stick to that decision; take a brief break after each episode to get out of the "auto play" loop; adhere to a consistent bedtime regardless of any unbearable cliff-hangers; and absolutely no streaming or other mobile devices in bed. To avoid inhaling unhealthy snacks, have healthy options—like fruits and vegetables—ready to go. And to make TV viewing a more communal activity, consider watching with a friend, family member, or romantic partner (perhaps put more emphasis on the *chill* part of the "Netflix & Chill" activity).

And of course we should all do our best to move. Get off your duff regularly, at least once an episode. Better yet, stream while exercising. If you are feeling ambitious, try implementing a 50/50 rule. For every episode you watch sitting or lying down, you must watch one doing some form of exercise.

Remember, the evidence suggests that using these strategies might help you enjoy the shows more. You may not earn any binge-racer, superfan street cred, but you will have a more positive and lasting memory of the experience.

10:00 PM—Check phone, again

Don't. Your phone should be put away so that it can't be a source of stress. You should be winding down. No email, text messaging, social media, or news feeds. You can find out about the latest political blunder while sitting on the toilet in the morning.

10:30 PM—Wash hair

The question "How often should you wash your hair?" was the eighth most popular "how" Google search in Canada for 2017. Clearly, people are really unsure about hair care. And this is something people deal with almost every day. It is a massive multibillion-dollar industry. So you'd think there would be lots of actual science out there to inform this decision. But I found almost nothing of value. There are a lot of opinions, mostly from people working in the industry (let's call them Big Shampoo shills), but very little actual data. There are studies on things like the effect of shampoo on cortisol levels in the hair (not much, in case you're wondering) and whether cannabis shampoo (that's a thing?) will lead to testing positive for cannabis (it might), and many review articles—some funded by Big Shampoo—that make vague recommendations about the frequency of hair washing that aren't supported by science or any clinical evidence.

And yet the industry presents shampoo as if it *is* science-informed. The advertising uses sciency-sounding words—*protein*, *nourishment*, *purifying*, *anti-residue*, etc.—and often features images of things like molecules, double helixes, and chemistry equations to convey the impression that science backs up the product. Don't be fooled. This is an evidence-free zone. A certain percentage of us accept that. A 2018 survey found that 69 percent of customers know that hair care advertisements are misleading, while 82 percent think this shouldn't be allowed. Over the years, these advertisements have attracted a response from government truth-in-advertising watchdogs. (You can't, for instance, tell people your shampoo will make hair 10 times stronger after a single use, as one marketing campaign did.) Still, remind yourself to be on guard for sciency-sounding, science-free hair care noise.

Despite the tremendously weak evidence base, I think I can safely suggest two things. First, you may be washing your hair too often. Surveys indicate that over 60 percent of us wash our hair every day or nearly every day. Stop doing that. While there is variation in hair types and reasons to wash (if you work in a coal mine, I get the desire to wash often), there is no need for most of us to shampoo daily. In fact, as noted in one of the few academic articles on shampoos, written by a professor of dermatology at Duke University, "Shampooing is actually more damaging to the hair shaft than beneficial." For most people, a couple of times a week is likely sufficient. And no, your hair will not look terrible. It may even look less dry and have more shine.

Second, you don't need to buy a fancy shampoo. And, holy cow, there are some fancy products out there, some costing hundreds of dollars for a single bottle. A product called Ten Voss is priced at $300 for a smallish, admittedly pretty cool-looking, container. But is it any better than a $5 bottle from the corner store? Probably not. All shampoos contain basically the same stuff, including a detergent to remove dirt and oil, a lathering agent to give you suds, and several other ingredients that give the shampoo a particular smell, look, and feel but have little to do with how effective it is.

In 2017, Dr. Laura Waters, a professor from the Department of Pharmacy at the University of Huddersfield, did a casual experiment with her graduate students. After washing the students' hair with a range of shampoos with drastically different prices, the research team looked at the hair under a microscope. They found that, regardless of price, thickness, texture, or look, all the shampoos cleaned the same. Dr. Waters noted that if people enjoy their brand, there is no reason to change. "If you just want clean hair, however, then price doesn't matter and swapping brands could save you a small fortune in the long run."

Here is an interesting side note to consider, one that further highlights the arbitrary pricing of hair products. Women pay more than men for what amounts to essentially the exact same product, a reality that is part of a broader product-pricing phenomenon that has been called the "pink tax." A study by the New York City Department of Consumer Affairs looked at a range of consumer products and found that gender-based price disparities can be significant. For example, women's shampoo costs 48 percent more than men's. While the scent and bottle might be different from the men's product, is it really worth that kind of price hike?

10:45 PM—Floss teeth

You know the drill. You show up at the dentist's office and the hygienist asks you about your flossing behavior. "It's pretty good," you lie.

Many of us say this because we have been told, for decades, that flossing is an essential part of oral health. Good people floss. We are told this by our dentists, dental associations, schools, the multibillion-dollar floss industry, and even governments. In my province, for example, Alberta Health tells me via its website to "floss at least once a day" as part of basic dental care.

Truth is, most people are really bad at flossing and we don't do it often. And let's be honest, we probably never will. Behavior change is tough, particularly when it comes to oral health. A 2018 survey of over 8,000 Americans concluded that only 32 percent floss daily. But this was self-report data, so my guess is that an objective assessment would result in a much lower compliance rate. Indeed, a 2017 survey of Americans put the number of daily flossers at just 16 percent! And according to a 2015 study by the American Academy of Periodontology, 27 percent admit to lying to their dentist about flossing. It also found that 36 percent hate flossing so much that that they would rather do something unpleasant like clean the toilet. Some people would even prefer to listen to nails on a chalkboard.

But the more shocking truth is that the science around flossing is far from definitive. There is very little good evidence to support the idea that we should all be concerned about regular flossing. "The evidence is pretty weak for any health benefit," Professor Robert Weyant told me. "If it is done well and really consistently, it can have a modest impact on gingivitis" (a common gum disease).

Dr. Weyant is the professor of dentistry from the University of Pittsburgh that we met when we brushed our teeth at the beginning of this hypothetical day. His lack of enthusiasm for flossing is informed by studies like the 2011 systematic review published by the prestigious Cochrane Collaboration that concluded there is only weak and "very unreliable evidence" that flossing is associated with a small reduction in plaque and that there is no good evidence that it helps prevent cavities. This conclusion is consistent with a later scientific review, concluded in 2015, which found that "available studies fail to demonstrate that flossing is generally effective in plaque removal." The research was felt to be so inadequate, in fact, that the U.S. government removed flossing from its 2015 official Dietary Guidelines, where it had been listed since 1979.

Since these comprehensive scientific reviews—and the media attention that surrounded them, a controversy called "flossgate" by those in the dental community—some studies have found a correlation between flossing and better oral health. But, as we have seen many times in this book, correlation is not causation. People who floss are also likely to take better care of their health more generally. Indeed, a 2018 survey of 8,000 Americans found that tobacco users were less likely to floss than people would didn't use tobacco. The study also found that "adults in the highest income group had higher odds of daily flossing compared with the lowest income group." Income and tobacco use alone could, independent of flossing, account for a lot of the differences in oral health.

It has also been suggested that if people flossed better, there would be more evidence to support flossing. (Which is a bit like saying if people ate more fruits and vegetables, people would eat more fruits and vegetables.) For example, an article titled "In Defense of Flossing" published in 2017 in the *Journal of Evidence-Based Dental Practice* argued that using better communication strategies—such as tailoring flossing advice to individual needs—would result in more and better flossing and a concomitant reduction in oral health problems. There are two challenges with this recommendation. First, generating population-level sustainable behavior change is really hard, especially when it comes to something that, for many, is less appealing than cleaning a toilet. Second, we still don't know that more and better flossing will result in health benefits.

"When it comes to flossing," Dr. Weyant told me, "you need to do it exactly right all the time to get any possible advantage. But most people don't do it right and then they stop." He sounded a bit frustrated with the futility of it all. "Very few people do it right. It is a skill." In some of the studies that have found a reduction in cavities, the flossing was done by professionals, such as dental hygienists. Getting the average person to floss with the same dexterity and diligence as a dental health professional is completely unrealistic. As one dentist wrote in the *Guardian* in 2016, "Even in controlled studies, after instruction, the patients taking part couldn't floss properly."

(In case you are wondering, here is a very rough guide to proper flossing. Gently go up and down between your teeth and gently curve the floss around the base of each tooth, including going below the gumline. Hold the floss against the tooth and rub the sides of each tooth. Use a clean section of floss for each tooth and never snap the floss into the gums.)

Some dental organizations give a grudging nod to the studies that found a lack of supporting evidence, but then still insist, paradoxically, that there is evidence. The Ontario Dental Association published a document called *Your Oral Health* in which they note that the flossing evidence "is still considered unreliable." But in the very next paragraph, they recommend flossing. To support this conclusion, they rely on anecdotal clinical reports, which, of course, is exactly what an evidence-based approach to health care should avoid. "In our clinical practice, we see evidence [that flossing works] every day," a dentist is quoted as saying. Good scientific studies are designed to cut through the biases inherent in the use of this kind of testimonial.

The American Dental Association, among others, now explicitly accepts the lack of good evidence to support flossing. But the organization still won't let flossing go off to the great spittoon in the sky. In 2016 it provided an endorsement of the practice that feels equal parts desperate and tepid: "While the average benefit is small and the quality of the evidence is very low (meaning the true average benefit could be higher or lower), given that periodontal disease is estimated to affect half of all Americans, even a small benefit may be helpful."

It will be a challenge to get the needed data to provide a definitive statement on the value of flossing. Running a large clinical trial wouldn't be easy or cheap. Also, the idea that flossing could help does have a degree of biological plausibility, even without solid clinical evidence. Allowing food to remain crammed in the spaces between your teeth can't be good. Flossing is a cheap intervention that, as far as I can tell, doesn't have a big downside. And flossing might make your teeth simply feel better. If you want to floss, go ahead and floss. There may be a small benefit, particularly if you are a flossing superstar and you do it well. (But you aren't, so you won't.)

I think it is fair for the dental community to use these kinds of points to push back against blanket statements that suggest that flossing has been proven not to work, as that hasn't been proven at all. Still, I remain skeptical about the value of flossing. If it worked well and was really essential to dental health, those benefits would be more apparent, even given the limitations of the existing research.

The underperformance of flossing should also be considered in the context of the underperformance of dental hygiene more broadly. Despite our heavy investment in dental care over the past few decades, a 2015 study found that "the global age-standardized prevalence and incidence of untreated caries remained static between 1990 and 2010." Pushing flossing and other forms of dental hygiene hasn't helped at all. If you think that sounds like a damning conclusion, consider this large 2018 systematic review that looked at the effect on reducing cavities of all forms of personal oral hygiene (defined as brushing teeth with or without the use of a range of flossing devices). The study, by a team from the University of Washington and Harvard, looked at all the available randomized trials and included information from almost 750 research participants. What it found was that "personal oral hygiene in the absence of fluorides has failed to show a benefit in terms of reducing the incidence of dental caries." When it comes to preventing cavities, it is mostly about delivering fluoride to our teeth. Given that conclusion, we should stop repeating messages that are not backed up by good science and put our limited resources and energy into the interventions that are supported by clear evidence.

Before we leave the flossing story and you hop into bed, one last bit of dental advice. If you are going to floss, do it before you brush. A 2018 study found that doing it in this order will "increase the fluoride concentration in interdental plaque." If you floss after you brush, you may be removing the cavity-fighting fluoride—a conclusion I take as yet more evidence that much of the benefit we derive from our oral health practices flows from fluoride and not from our aptitude with string and bristles.

10:50 PM—Sex

Sure. Why not? Couple. Solo. Whatever.

Sex is healthy, fun, relaxing, and almost always a fine decision (with all the usual consenting-adults caveats). Sex, and a positive attitude toward sex, is good for relationships and our overall well-being. At the same time, sex isn't a contest. You aren't competing with other couples—or with the inaccurate pop culture and porn portrayals of sex. You don't get points for frequency or acrobatic panache.

But many of us do worry about these things. We think other people are having much more sex than they actually are. A 2018 survey of American and U.K. adults found that men estimate that others have four times more sex than they actually do. Women's predictions were better, but they still guessed that other women had sex twice as often as they actually do. It seems that many of us believe everyone is involved in some wild orgy club and we never got the invitation.

Unfortunately, these misperceptions matter. People interpret their own sexual satisfaction by comparing themselves with what they believe others are up to. A 2014 study, for example, found that, in general, the amount of sex people have is positively associated with happiness, but that the amount of sex we believe *others* are having is negatively associated with our happiness. In other words, we don't like it when we think our friends and neighbors are having more sex than we are.

We should stop this no-win social comparison game. The right amount of sex, at least for a couple, is the amount that makes them feel happy, close, and satisfied. It is the amount that is right for them. And more isn't always better. Although studies have consistently found that frequency is correlated with a couple's satisfaction with their relationship (lots going on with that correlation, including the fact that more sex is likely a marker for a good relationship), after about once a week there is no improvement in well-being. A 2015 study randomly assigned a group of couples to double their frequency of intercourse. But this coupling doubling did not lead to more happiness and actually decreased their enjoyment of the sex. The researchers speculate that this is because doing it more (even in the name of science!) took away from the spontaneous and intrinsic motivations for intimacy.

10:55 PM—Cuddle

Yes, this is five minutes later. The average married heterosexual couple has sex for just over five minutes. (Seriously? Five minutes? On the plus side, I suppose that leaves more time for sleep.) When you are finished, please cuddle. Studies have found that cuddling has many benefits. A 2014 study of over 300 Canadian couples found that the "duration of post sex affection was associated with higher sexual satisfaction and, in turn, higher relationship satisfaction."

11:00 PM—Sleep

Here we are at the end of a long day filled with dozens of decisions. You must be exhausted! You need a good night's sleep. So, dammit, make the decision to get a good night's sleep!

As I have personally experienced, this is frequently easier said than done. I'm a terrible sleeper. I fall asleep fine, but often wake in the middle of the night. I'm not alone in this predicament. A large portion of the adult population has trouble either falling asleep or staying asleep. And poor sleeping habits are associated with a range of issues, including obesity, Alzheimer's, cardiovascular disease, depression, type 2 diabetes, poor academic performance, a decrease in productivity, and workplace and motor vehicle accidents. As noted by the Division of Sleep Medicine at Harvard University, studies have shown that "sleeping five hours or less per night increased mortality risk from all causes by roughly 15 percent."

I won't hammer away at the reality that we all need to sleep more. You've probably heard that from everyone from Arianna Huffington to Michelle Obama to Beyoncé. The fact that many of us don't get enough sleep is driving a massive and growing sleep industry. For decades, sleep was viewed as something that weak people did. It was a colossal waste of valuable time. A lack of sleep was a badge of honor. It was a sign that you were ambitious and willing to do anything to succeed. You can sleep when you're dead and all that. Now, though, there is a broad awareness that these ideas are not only wrong from a productivity point of view—which was, ironically, the reason people were staying awake—but that they're also downright harmful to both individuals and society. The script has completely flipped. Very few now buy into the you-snooze-you-lose cliché. This is a time when professional sports teams hire sleep coaches, celebrities brag about their sleep routines on social media, and LeBron James's personal trainer declares that "sleep is far and away his most valuable asset." (Really? More than LeBron's six-eight frame, freaky skill, and competitive nature?) You don't need more convincing from me. Sleep is important.

I also think most of us are now aware of the basic strategies for improving our sleep. In general, adults should strive for between seven and eight hours a night (with variations largely due to our biologically determined chronotypes). There is no magic formula, but most experts agree that a good night's sleep will be aided by a consistent bedtime routine that includes going to bed at roughly the same time each night, and avoiding stimulants like caffeine and the consumption of too much alcohol near bedtime. Your bedroom should be sleep-friendly—quiet, dark, cool in temperature, and, of course, with a cozy, clean bed. And the bedroom should be reserved for just sleep and sex. As noted by the U.K.'s National Health Service, "Unlike most vigorous physical activity, sex makes us sleepy." (I like how they optimistically portray sex as "vigorous," particularly given the underwhelming five-minute duration referred to above.)

So yes to sex but a *hard no* to TVs, tablets, and smartphones. Stop bingeing TV and catching up on social media while in the bedroom. A fairly strong body of evidence is starting to emerge that the blue light emitted by these devices disrupts sleep. There is some evidence—though not much quality research, so more science is needed before definitive claims can be made—that blue-light filters, such as amber-tinted glasses, may help. More problematically, however, these devices are portals to our work world, news feeds, and other sources of stress that are not conducive to a wind-down state of mind. Light filters won't help with this problem. Online engagement keeps the mind active and removed from the sleep environment. Even if you are not looking at your work email or social media, simply holding the phone is a subtle reminder of these areas of your life. Just being around your phone can raise levels of the stress hormone cortisol, adding to feelings of anxiety. A 2017 study, for example, found that social media use within 30 minutes of bedtime was "associated with disturbed sleep among young adults." And a 2018 study from the University of East London found that "sleeping without smartphones improves sleep, relationships, focus and wellbeing."

Still not convinced? Some studies suggest that smartphones in the bedroom are killing our sex life. People are having less sex than in past decades. While this is undoubtedly a complex issue, some researchers believe that technology is at least partly to blame. A 2018 study from the U.K. found that peaks in internet data use appear to happen late at night, reflecting the use of social media and online entertainment after we all crawl into bed. David Spiegelhalter, a psychologist at the University of Cambridge, has suggested that couples are now less interested in sex because of this connectivity. Indeed, a 2018 survey of 2,000 American adults found that people in relationships spend an average of 40 minutes on their phone before going to bed and that 93 percent leave their cellphone within reach while they sleep. Cellphone use (and not sex) is the number one activity during the last hour before sleep. More than half of the couples surveyed felt that cellphones in the bedroom were causing them to miss out on quality time with each other. And 25 percent see their smartphone—not their partner—last before they close their eyes.

Put away the phone!

Sleep is having a cultural moment. And that is, mostly, a good thing. Shut-eye has been neglected for far too long. But with the enthusiasm for all things sleep has come a tidal wave of sleep products, books, advice columns, supplements, and sleep experts and gurus with varying degrees of credibility and knowledge. Sleep has become something we are now supposed to work hard at, as if we were training for a marathon or preparing for a solo at Carnegie Hall. Never mind that sleep is something humans have done naturally for *our entire existence*. Our bodies know how to sleep. It is wired into our biology. But now we are sold gadgets that are supposed to allow us to do it better. And we're falling for them. The sleep industry is projected to soon be worth $100 billion.

We don't need to try so hard. We need to relax about relaxing. But all this talk about sleep has created a sleep effort paradox. Studies have found that just being anxious about sleep—which is what the growing sleep industry asks us to do so that they can sell us stuff to help with our sleep anxiety—may actually make it more difficult to sleep. There is so much pop culture noise about sleeping that people may start obsessing not just about getting more of it but about how to do it correctly. This may, in turn, lead to even more sleep problems. In a 2017 paper published in the *Journal of Clinical Sleep Medicine*, the authors labeled the inappropriate fixation on sleep "orthosomnia" (meaning "correct sleep"), a situation whereby a "perfectionistic quest for the ideal sleep" leads to an unhealthy fixation, which leads to more worry and, as a result, less sleep.

The growing popularity of sleep tracker technology is a good example of the issues associated with the new sleep culture. I've tried a variety of these devices. I've worn a special sensor on my head that was supposed to pick up brain waves. I slept on a mat designed to pick up my nocturnal movements. And I tried the kind worn on the wrist, designed to monitor the various stages of my sleep. In the morning I got feedback about various aspects of my sleep, presented with impressive and sciency-looking graphs and pie charts. What, exactly, is a person supposed to do with this information? If I have a bad score, do I try to sleep harder the next night? If I feel refreshed in the morning but the tracker said I had a bad sleep, should I revise my sense of being rested? More problematic, each device gave me a different response. One told me I'd earned a sleep "score" of 73 percent—which I think is supposed to be good—and another awarded a less impressive grade.

The evidence regarding the accuracy of such devices is, to put it kindly, very mixed. A 2017 review of the research on consumer sleep trackers concluded: "These devices have major shortcomings and limited utility, as they have not been thoroughly evaluated in clinical populations." Part of the problem is that the trackers don't do a good job differentiating between different states of sleep. For example, lying still while watching a show on your phone may register as light sleep, which it most definitely isn't.

I also think it is ironic that we are turning to a tech device that is often paired with our phone in order to get a better night's sleep. Our phones and other technology are one of the key reasons we are up half the night. It feels a bit like fighting a weed infestation by planting dandelions.

But for me, the bigger issue with sleep trackers is that they intensify the sleep effort paradox. Because these devices provide a measurement—which, as we have seen, might be largely meaningless—they invite us to compare successive nights of sleep and, as a result, put more and more effort into sleeping longer and better. The quantification of sleep allows us to treat it as a competition. This is, of course, exactly the wrong way to think about sleep, which should be governed by a more natural rhythm and a less intense vibe. Sleep isn't a race that can be won.

I am happy we are taking sleep more seriously. But as so often happens with wellness trends, a straightforward activity is made to seem more complicated in order to facilitate the growth of an industry. The strategy that has helped me the most with my sleep issues is one that fits nicely with the underlying theme of this entire book. Relax. Chill. Unplug. Wind down. Do what feels right for you. Don't let the sleep industry sell you on the idea that you need special pillows, mattresses, blankets, supplements, or a highly regimented and meticulously monitored routine. Stick to the basics, including a consistent sleep and wake time, a cozy sleep environment, and the avoidance of technology, caffeine, too much alcohol, and overeating before bed. You don't need to quantify every moment of your life, especially your sleep. Indeed, doing that could backfire.

THE YOUR DAY, YOUR WAY RULES

Late at night when I'm trying to sleep, I'm often pulled back to the problems that dogged me during the day. I'll rehash the issues, trying to work my way to a solution. It is a bad habit, one that is not conducive to slumber.

This book has covered a busy hypothetical day. I've examined controversial topics that are confused by layers of misinformation and, sometimes, ideological spin that too often lead us down the wrong path or stress us out unnecessarily. I've talked about the decisions we could make, but often don't, that may be beneficial.

But the core of this book is really about the facts behind our daily decisions and how and why those facts are misrepresented, misinterpreted, or ignored. While writing this book I *did* have many sleepless nights as I tried to work through how, in this age of unprecedented levels of access to knowledge, we so frequently and consistently get it wrong. I've thought about this for decades. As I said at the beginning of this book, everyone wants to make decisions that are right for them and no one wants to be misled.

A 2018 study by the Knight Foundation found that only 27 percent of Americans say they are "very confident that they can tell when a news source is reporting factual news" and that 58 percent feel that it is "harder rather than easier to be informed today due to the plethora of information and news sources available." Another large international survey, the 2018 Edelman Trust Barometer, found that 59 percent of the public are "not sure what is true and what is not."

As such statistics show, when it comes to using information to make our day-to-day decisions, these are crazy and challenging times. But why is misinformation having such a huge impact on our lives? How did we get to this place, and what can we do about it?

As I was writing this book I had the opportunity to chair a panel discussion on science denial hosted by the New York Academy of Sciences and Rutgers Global Health Institute. Our session was focused on these exact questions. Specifically, we were tasked with exploring the social forces that have facilitated the spread of misinformation and what we can do to fight back, both as individuals and as a society. The panel included five stellar scholars from institutions around the world. We debated the role of various factors, such as public trust, social media, narratives, and celebrity culture. But for me, Kelly M. Greenhill, a professor with Tufts University and Harvard's Kennedy School for Science and International Affairs, provided the most concise summary—six social trends that go a long way toward explaining why this is the era of misinformation. Below is my take on Kelly's list. The analysis is mine, but Kelly's list is a useful structure.

High levels of generalized public anxiety—Fear is everywhere. We saw it with our concerns about milk, fluoride, the dangers of bike commuting, and whether we should let our kids walk to school. It plays a role in our guilt about how much time we spend with our kids and how we think about handshakes, hugs, and toilet seats. We know that, unfortunately, fear works—it helps to sell products and ideas. Playing to our fear feeds our hardwired cognitive bias to see and remember dramatic events, such as a bear attack, a child abduction, or a skydiving accident. This skews our thinking, making it difficult to reach dispassionate, rational, and evidence-informed decisions.

Historically low levels of trust—Research tells us that public trust in our basic institutions is at a near-record low. Gallup data from the U.S. shows that in 1975, fully 80 percent of the public had a great deal or quite a lot of confidence in the medical system. Today that number is 30 percent. Confidence in newspapers dropped from a high in 1979 of 51 percent to 23 percent today. I could go on. A breakdown in trust can cause people to turn away from good science because they don't trust the source of the information—be it due to concern about the impact of the profit motive or the role of political agendas. This breakdown in trust makes room for conspiracy theories and alternative, evidence-free explanations. When people can't make sense of the world and don't trust public institutions, it is easier for misinformation to flourish. It is easier to believe that raw milk is safe, fluoride is a mind control drug, and Big Pharma is withholding the truth about the value of vitamins and supplements. Indeed, as noted in this book, just being exposed to conspiracy theories can confuse public discourse and influence your beliefs and intentions, even if you aren't a hard-core conspiracy theorist.

High levels of polarized public discourse—There no longer seems to be a middle ground on anything. Whether it is a debate about climate change or the value of a gluten-free diet, or a rant about ranting, the public representations are, more often than not, almost always extreme. You now must be 100 percent for or 100 percent against. No equivocating allowed. A 2018 study from MIT produced visual representations of public discourse on social media. What it looks like is an almost perfect U shape. The extremes are loud and frequently heard; the middle near silent. The phenomenon of false balance in the media feeds this polarization by presenting the science relevant to social controversies in an inaccurate and misleading manner. False balance happens when a media story suggests that there are two equal or near-equal sides to a story when, in fact, the weight of the evidence clearly favors one side. This can happen when one side of the debate—often those driven by fear or a particular ideological agenda—is more motivated to engage the topic than the other side. A 2018 study from Canada on the media coverage of the debate about the fluoridation of water is a good example of how this can play out. The researchers found that media stories misrepresented the science in an unbalanced manner, largely because those against fluoride were more motivated and better organized to speak out than those in support of the public health initiative. This led to news coverage that was "framed in terms of risks instead of benefits." It creates a polarized, two-sides-battling narrative. This storytelling strategy is good for selling papers and getting clicks, but not so good for public discourse and policy development.

The emergence of disruptive communication technologies—It wasn't long ago that the sources of information were fairly restricted. There were a handful of TV channels, a limited number of radio options, local newspapers, magazines, and books. These media outlets controlled and filtered what we heard about any given topic. With the rise of the internet and social media, how information is shared has been forever altered. Much of this is wonderful—allowing for unique forms of entertainment and artistic expression, more access to science, and new online communities. But, as we have seen throughout this book, it has also allowed less than ideally informed ideas not only to survive, but to thrive. Studies have shown that bad news, lies, and misinformation travel faster and further on the internet. This has fed the pop culture noise that now overwhelms public discourse, and it makes it near impossible to tease out the actual science from the nonsense. Social media platforms both facilitate the growth of polarized discourse and increase the sway of confirmation bias—the tendency to find information that confirms our preconceived ideas.

The rise of new information gatekeepers—Increasingly, platforms like Google, Facebook, and Twitter influence what gets our attention and what doesn't. They influence what gets covered in the traditional news outlets and which academic publications get public attention. The result is a dynamic where the editors' and journalists' ability to shape public discourse is overwhelmed, for better or worse, by users' clicks and social media shares—which, in turn, likely influences future editorial content. A small, preliminary, or poorly done study about the power of napping or the nutritional value of coffee may circulate widely and quickly on Twitter, allowing it to become part of the public consciousness much more rapidly than in the past. And once the story is out there, it can be hard to retract or debunk.

Perhaps more challenging is the impact of search engine results. What appears on the first page of a Google search, for example, is not, obviously, a list of the most reliable, science-informed information related to a particular query. If you search almost any topic covered in this book, the first page of the search results will be a mash-up of some good stuff from trustworthy sources and a bunch of completely wrong and potentially dangerous crap.

As I was writing this conclusion I googled "raw milk benefits." The top response—that is, the featured snippet at the top of the page—was a quote from one of the best-known purveyors of pseudoscience and health misinformation, Dr. Joe Mercola. This is an individual who has received warnings from the FDA to stop making unfounded claims, has been sanctioned by the Federal Trade Commission for selling an unproven product, and has received an "F" from the Better Business Bureau. When Mercola tells you, as his website does, that pasteurized milk "can be detrimental to your health," you can assume this advice should be vigorously ignored. Yet this was the *featured* result on Google! It was the snippet, which is often the only thing people read. It is what Google Home or the Amazon Alexa would read out loud if you wanted an audio response. Out of the other nine results, there were only two with good, balanced, science-informed information about raw milk. Most of the other results took me to websites, including that of a group lobbying for the sale of raw milk, which had information that was horribly misleading. And remember, the first page of search results receives 95 percent of web traffic on that topic. The top listing—in this case Mercola's bunk-o-rama advice—receives by far the most traffic. Search results shape what we know about the world. And when it comes to informing decisions, if it isn't on the first page, it might as well not exist.

The presence of many actors willing to circulate bunk, hype, and fear—For me, this is perhaps the most frustrating element. A growing number of individuals, institutions, and even governments seem willing to actively use the new media platforms to manipulate our information environment. Much has been written about the most dramatic examples, such as government interference in the democratic process. At times this has reached far beyond politics, as when Russian Twitter trolls waded into the debate about vaccine safety as a way to create more general angst, polarization, and discord among the American people. (The trolls used messages like, "Don't get #vaccines. Illuminati are behind it.") Celebrities, conspiracy theorists, social media pontificators, and marketers pollute the internet with misleading misinformation and outright falsehoods. Although much of the information is absurd (the vile website NaturalNews had a headline that declared, "Measles Vaccines Kill More People Than Measles, CDC Data Proves"), the noise still can have an impact. Merely being exposed to this nonsense can influence our beliefs and, ultimately, our decision-making.

In addition to these obvious sources of science-free nonsense, we need to remember that legitimate scientific institutions can also add to the noise that makes decisions difficult. Researchers, universities, and funding entities unnecessarily hype and twist their research so that it will get traction in both the traditional news outlets and on social media. Overly enthusiastic representations of research can confuse public discourse about complex topics. More often than not, either this science hype portrays the science as more definitive than it really is or the media representations of the research downplay the limits of the relevant methodologies.

I could go on. But I think this list of six key social trends is enough to make the central point: we live in a society that is now engineered to create confusion, stir anxiety, and spread misinformation. It is no wonder that making a decision can be so challenging.

So what can we do? How should we respond? Here are six strategies that will allow both you and me to relax and put this decision-filled day behind us so that we can get a good night's sleep. (Incidentally, research has demonstrated that writing a to-do list is a good way to wind down and promote a restful night. So let's get to it!)

Arm yourself with tools that will help you recognize misinformation—Sorting through all the pseudoscience, fake news, conflicting data, and science hype can be a huge challenge even for people like me who do this as part of their job. But here are a few strategies that can help.

First, consider the kind of science used to support the claim. If it is an animal study, has a small sample size, or finds only an association (repeat after me: correlation is not causation!), treat the research as preliminary or investigative.

Second, always consider the body of evidence on any given topic. It is extremely rare that a single study, even if large and well done, will reshape the scientific consensus. (Ignore headlines that say otherwise.)

Third, do not be persuaded by anecdotes or testimonials. Although they can be compelling and may invite further inquiry, they are not good evidence.

Fourth, consider the source (who is the author?) and be suspicious of any claim that is supported by a conspiracy theory or claim to secret information. If stuff worked well, we'd know. Promise.

Fifth, consider biases and conflicts. We all have them, but for some individuals and entities a particular point of view represents their entire agenda (think of websites trying to legalize the sale of raw milk or scare you about vaccines) or a path toward profit. Such biases can have a profound influence on how science is interpreted and presented.

Sixth, be wary of vague phrases ("increase energy levels," "detoxify body," "boost immune system") and overpromises ("revolutionary," "game changer," "guaranteed weight loss").

Seventh, be alert to "hot stuff" bias, which is the tendency for fashionable science topics to get more attention and be subject to less critical reflection.

Finally, remember that we all have cognitive biases—especially our tendency to see things that confirm what we already believe—that can change our interpretation of the evidence. Challenge yourself. Stay curious. And consider different perspectives.

Relax. A bit of critical thinking can help you cut through the noise.

Don't let fear rule your life—As we saw again and again in this book, we often overestimate risks. Or we worry about the wrong kind of risk—such as the remote risk of child abduction instead of the well-known risks associated with physical inactivity. This happens for a number of reasons, including the fact that dramatic events are easier to imagine. If you find that a particular decision is being dominated by fear, do your best to gain an understanding of the actual magnitude of the risks. It won't always change your mind—I'm still afraid of bears and remain something of a germophobe—but at least it may lower your anxiety level. Remember that in most of the world, this is the safest time to be alive, whether you are a child, teenager, or adult. Ignore the fearmongers.

Relax. The risk probably isn't as bad as you think it is.

Look to the body of evidence and recognize that science is often uncertain—Whether you are deciding about standing desks, breakfast, flossing, or how to improve your sleep, the research is often much more equivocal than portrayed in the popular press. Too often, the next new health- or wellness-improving fad is presented as an essential and straightforward fix for what is actually a complex, nuanced problem. Some decisions are trivial—the five-second rule and the toilet seat debate—but still induce strong reactions based largely on science-free pontificating. Other times, ideas have intuitive appeal, such as the conventional (and wrong) wisdom around waking up early or venting our rage. These kinds of misrepresentations and beliefs can harm. They can also lead us to waste time and money and distract us from the basic, science-informed strategies that do work.

Relax. There are no magical solutions. (Except maybe exercise.)

Don't get fooled by the illusion of difference—We make many of the decisions we do because we think there are quantifiable differences between things. We think we are choosing something that is objectively better, such as when we buy expensive wine or a fancy toothbrush. If you enjoy this selection process or, perhaps, the history and culture that informs these decisions, terrific. But recognize that, for many things, there is probably little meaningful difference. (But don't take away my overpriced espresso! Worth every cent.)

Relax. The difference probably isn't that meaningful. Save your money.

Focus on the fundamentals and ignore the "wellness noise"—Much of the misinformation discussed in this book touches on how to maximize health and safety for both individuals and families. Despite the multitrillion-dollar wellness industry that seeks to sell us supplements, high-tech monitoring devices, bottled water, extreme diets, and sleep aids, we have known the science-informed basics of a healthy lifestyle for decades: don't smoke, exercise, eat real food, maintain a healthy weight, drink alcohol not at all or in moderation, sleep, and have good relationships. We can also take logical preventive steps, like wearing a seat belt, getting vaccinated, and washing our hands. For most humans, these well-known steps are all that is required. Everything else is just fiddling at the margins. Of course, socioeconomics also play a huge—perhaps dominant—role. Indeed, if we are looking to improve the health of our communities, effecting social and environmental changes that focus on these basics is what is likely to have the biggest impact.

Relax. Find a sustainable, enjoyable lifestyle that embraces the basics.

Finally, don't ignore the science that will help you relax!—Much of this book has been about tackling the misrepresentations of science that cause anxiety or poor decisions, but don't overlook the research about the steps we can take to lead a more relaxed life. For example, the body of evidence regarding the detrimental effect of smartphones on sleep and interpersonal interactions is growing. And there is emerging evidence about how we should manage and think about our work time, home activities, and connectivity, including managing email, social media, and streaming entertainment. Any effort to make our days less fragmented, frantic, and stress-inducing is worth considering.

Relax. And turn off your damn phone.

Zzzz.

ACKNOWLEDGMENTS

This book covers a lot of different topics. I sought advice from many experts throughout the world. I did this in person, over the phone, and via email. (For the sake of brevity, I did not differentiate in the text of the book.) These people helped me navigate a vast literature and provided invaluable insight and context. I am tremendously grateful for their input and time. Thank you: Satchidananda Panda, Salk Institute; Tanis Fenton, University of Calgary; Lenore Skenazy, Free Range Kids; Geertrui Van Overwalle, Leuven University; Kay Teschke, University of British Columbia; Rebecca Puhl, University of Connecticut; Jayne Thirsk, Dietitians of Canada; Rhonda Bell, University of Alberta; Kim Raine, University of Alberta; Andrew Velkey, Christopher Newport University; Joan Salge Blake, Boston University; James McCormack, University of British Columbia; The Angry Chef; Ellen Macfarlane Gregg, University of Waterloo; Daniel Flanders, Kindercare Pediatrics; Brett Finlay, University of British Columbia; Leah McGrath, Ingles Markets; Paul Kelley, Open University; Charles Samuels, University of Calgary; Grant Ritchie, dentist; Raya Muttarak, Wittgenstein Centre for Demography and Global Human Capital; Robert Weyant, University of Pittsburgh; David Allison, Indiana University; James A. Betts, University of Bath; Bob Burden, Serecon; James C. Coyne, University of Pennsylvania; Steven Hoffman, York University; Elizabeth Mostofsky, Harvard University; Alan Levinovitz, James Madison University; Jen Gunter; Ryan C. Martin, University of Wisconsin–Green Bay; Gary

Schwitzer, HealthNewsReview, University of Minnesota; Rachel McQueen, University of Alberta; Derreck Kayongo, Global Soap Project; Colleen Carney, Ryerson University; Thomas Hills, University of Warwick; Vanessa Patrick, University of Houston; Emma Russell, Kingston University London; Birsen Donmez, University of Toronto; Donald Schaffner, Rutgers University; Kelly M. Greenhill, Harvard; Stuart M. Phillips, McMaster University; Jeff Vallance, Athabasca University; Josephine Chau, Macquarie University; Silvia Bellezza, Columbia University; Oriel Sullivan, University of Oxford; Charlene Elliott, University of Calgary; Stanley Kutcher, Dalhousie University; Yoni Freedhoff, University of Ottawa; Kostadin Kushlev, Georgetown University; Sara Kirk, Dalhousie University; Vinay Prasad, Oregon Health and Science University; Donald Vance Smith, Princeton University; Paula Fomby, University of Michigan; Melissa A. Milkie, University of Toronto; Lin Yang, Cancer Epidemiology and Prevention Research, Alberta Health Services; Lynn Barendsen, the Family Dinner Project, Harvard University; Kirk Barber, University of Calgary; Matthew Nisbet, Northeastern University; Jonathan Jarry, McGill University; and anyone I may have forgotten!

While writing this book I also participated in many relevant conferences and workshops, including a few that I helped to organize. In addition, I worked on my TV documentary, *A User's Guide to Cheating Death*, which gave me the opportunity to discuss these topics with a range of experts. I would like to thank them all, too numerous to mention, and my friends at Peacock Alley Entertainment for their patience and help.

I also appreciate the continued support from my colleagues at the University of Alberta and the Health Law Institute (many of whom had to tolerate hallway arguments about the topics in this book), including Ubaka Ogbogu, Blake Murdoch, Sandro Marcon, and Erin Nelson. I am particularly thankful for the phenomenal work done by Robyn Hyde-Lay. She helped with background research and had to put up with my theorizing about the direction of almost every section in the book.

I am grateful for the funding I receive for my research, including, *inter alia*, support from the Canada Research Chairs Program, the Pierre Elliott Trudeau Foundation, the Canadian Institutes of Health Research, the Network of Centres of Excellence program, and Genome Canada.

My agent, Chris Bucci, has been, as usual, amazing, providing constant support and nudging me to whip this concept into a book. Much credit and thanks must also go to Diane Turbide and the wonderful team at Penguin Random House Canada. I'd also like to give Diane a special shout-out for helping (over a "moderate" amount of drinks in Toronto) to come up with the idea for structuring this book around a hypothetical day. And a big thanks to Shaun Oakey, Justin Stoller, Alanna McMullen, and Lauren Park for their editorial work.

Of course, I save my biggest thanks for my wonderful and always supportive family, with whom I've spent many magical days.

REFERENCES AND NOTES

I did my best to include enough information in the text so that the relevant study could be identified in the reference list. For some of the references I have also included a key quote or summation. I hope that you find these interesting and useful. Although much of the work I referenced in the text is included below, this is not a comprehensive list of the relevant research. The goal is to provide a sense of both the academic literature and some of the popular commentary on each topic.

Introduction: Decisions, Decisions . . .

Agel, J, et al. "A 7-year review of men's and women's ice hockey injuries in the NCAA" (Oct 2010) 53(5) Canadian Journal of Surgery 319–323—"18.69/1000 athlete-exposures."

Berge, LI, et al. "Health anxiety and risk of ischaemic heart disease: A prospective cohort study linking the Hordaland Health Study (HUSK) with the Cardiovascular Diseases in Norway (CVDNOR) project" (1 Nov 2016) 6(11) BMJ Open e012914—Health anxiety associated with increased health risks.

Boon, S. "21st century science overload" (7 Jan 2017) Canadian Science Publishing—"Approximately 2.5 million new scientific papers are published each year."

Curran, T, et al. "Perfectionism is increasing over time: A meta-analysis of birth cohort differences from 1989 to 2016" (28 Dec 2017) 145(4) Psychological Bulletin.

Dyson, T. "Heart attacks less frequent, less deadly since 1990s" (15 Mar 2019) UPI.

"Global life expectancy up 5.5 years since 2000: WHO" (4 Apr 2019) Medical Xpress.

Gramlich, J. "5 facts about crime in the U.S." (3 Jan 2019) PEW Research Center.

Herrero, S, et al. "Fatal attacks by American black bear on people: 1900–2009" (11 May 2011) 75(3) Journal of Wildlife Management 596–603.

Hirshleifer, D, et al. "Decision fatigue and heuristic analyst forecasts" (Feb 2018) National Bureau of Economic Research, NBER Working Papers 24293—"Forecast accuracy declines over the course of a day as the number of forecasts the analyst has already issued increases."

Ingraham, C. "There's never been a safer time to be a kid in America" (14 Apr 2015) Washington Post—Data on safety of kids, including how the child pedestrian rate of death has declined steadily.

Jagiello, RD, et al. "Bad news has wings: Dread risk mediates social amplification in risk communication" (29 May 2018) 38(9) Risk Analysis.

Jinha, AE. "Article 50 million: An estimate of the number of scholarly articles in existence" (Jul 2010) 23(3) Learned Publishing 258–263—"From the first model of the modern journal, *Le Journal des Sçavans*, published in France in 1665, followed by *Philosophical Transactions* published by the Royal Society in London later that year, the number of active scholarly journal titles has increased steadily."

Kabat, GC. *Getting Risk Right* (2016) Columbia University Press.

Lawler, EE, et al. "Job choice and post decision dissonance" (Feb 1975) 13(1) Organizational Behavior and Human Performance 133–145.

Limburg, K, et al. "The relationship between perfectionism and psychopathology: A meta-analysis" (Oct 2017) 73(10) Journal of Clinical Psychology 1301–1326.

Linder, JA, et al. "Time of day and the decision to prescribe antibiotics" (Dec 2014) 174(12) JAMA Internal Medicine 2030–2031.

Luu, L, et al. "Post-decision biases reveal a self-consistency principle in perceptual inference" (Jul 2018) eLife.

Nania, R. "Teen drug use at 'all-time lows': How to keep it there" (25 Jan 2018) WTOP.

National Cancer Institute, "Cancer statistics" (27 Apr 2018)—"In the United States, the overall cancer death rate has declined since the early 1990s."

National Institute on Drug Abuse, "Monitoring the future 2017 survey results" (Dec 2017).

Pachur, T, et al. "How do people judge risks: Availability heuristic, affect heuristic, or both?" (Sep 2012) 18(3) Journal of Experimental Psychology: Applied 314–330.

Pennycook, G, et al. "Prior exposure increases perceived accuracy of fake news" (2018) 147(12) Journal of Experimental Psychology: General.

Rajanala, S, et al. "Selfies—Living in the era of filtered photographs" (Dec 2018) 20(6) JAMA Facial Plastic Surgery—"These apps allow one to alter his or her appearance in an instant and conform to an unrealistic and often unattainable standard of beauty."

Rosenfeld, P, et al. "Decision making: A demonstration of the postdecision dissonance effect" (30 Jun 2010) 126(5) Journal of Social Psychology 663–665.

Roser, M. "Most of us are wrong about how the world has changed (especially those who are pessimistic about the future)" (27 Jul 2018) Our World in Data—Nice summary stats on disconnect between perception and reality on global issues like poverty and life expectancy. See also Roser, M, "Memorizing these three statistics will help you understand the world" (26 Jun 2018) Gates Notes—"Since 1960, child deaths have plummeted from 20 million a year to 6 million a year."

Schaper, D. "Record number of miles driven in U.S. last year" (21 Feb 2017) NPR—Americans drove "a record 3.22 trillion miles on the nation's roads last year, up 2.8 percent from 3.1 trillion miles in 2015."

Smith, MM, et al. "The perniciousness of perfectionism: A meta-analytic review of the perfectionism-suicide relationship" (Jun 2018) 86(3) Journal of Personality 522–542.

Tierney, J. "Do you suffer from decision fatigue?" (17 Aug 2011) New York Times Magazine.

Van Noorden, R. "Global scientific output doubles every nine years" (7 May 2014) Nature.

Vohs, KD, et al. "Making choices impairs subsequent self-control: A limited-resource account of decision making, self-regulation, and active initiative" (Aug 2014) 1(S) Motivation Science.

Westermann, RW, et al. "Evaluation of men's and women's gymnastics injuries: A 10-year observational study" (Mar 2015) 7(2) Sports Health 161–165.

PART I: MORNING

Wake up!

Ackermann, K, et al. "The internet as quantitative social science platform: Insights from a trillion observations" (Jan 2017)—Wake times around the world.

Åkerstedt, T, et al. "Sleep duration and mortality—Does weekend sleep matter?" (Feb 2019) 28(1) Journal of Sleep Research—"Short, but not long, weekend sleep was associated with an increased mortality in subjects <65 years."

Bowers, JM, et al. "Effects of school start time on students' sleep duration, daytime sleepiness, and attendance: A meta-analysis" (Dec 2017) 3(6) Sleep Health 423–431—"Later starting school times are associated with longer sleep durations. Additionally, later start times were associated with less daytime sleepiness (7 studies) and tardiness to school (3 studies) . . . Overall, this systematic analysis of SST studies suggests that delaying SST is associated with benefits for students' sleep and, thus, their general well-being."

Duncan, MJ, et al. "Greater bed- and wake-time variability is associated with less healthy lifestyle behaviors: A cross-sectional study" (Feb 2016) 24(1) Journal of Public Health 31–40—"Greater bed-time variability is associated with a less healthy pattern of lifestyle behaviors. Greater consistency in sleep timing may contribute to, or be reflective of, a healthier lifestyle."

Hershner, S, et al. "The impact of a randomized sleep education intervention for college students" (15 Mar 2018) 14(3) Journal of Clinical Sleep Medicine 337–347—Study found that sleep can improve performance.

Ingraham, C. "Letting teens sleep in would save the country roughly $9 billion a year" (1 Sep 2018) Washington Post.

Jankowski, KS. "Morningness/eveningness and satisfaction with life in a Polish sample" (Jul 2012) 29(6) Chronobiology International 780–785—Morningness related to greater life satisfaction.

Kalmbach, DA, et al. "Genetic basis of chronotype in humans: Insights from three landmark GWAS" (1 Feb 2017) 40(3) Sleep—"Taken together, heritability estimates suggest that genetic factors explain a considerable proportion, up to 50%."

Kelley, P, et al. "Is 8:30 a.m. still too early to start school? A 10:00 a.m. school start time improves health and performance of students aged 13–16" (8 Dec 2017) Frontiers in Human Neuroscience—"Changing to a 10:00 a.m. high school start time can greatly reduce illness and improve academic performance."

Knutson, KL, et al. "Associations between chronotype, morbidity and mortality in the UK Biobank cohort" (Aug 2018) 35(8) Chronobiology International—Later timing of sleep associated with increased health risks; study that found 27.1 percent of the population are definite morning types, 35.5 percent are moderate morning types, 28.5 percent are moderate evening types, and 9 percent are definite evening types.

Lee, CJ, et al. "Law-based arguments and messages to advocate for later school start time policies in the United States" (Dec 2017) 3(6) Sleep Health 486–497.

Manber, R, et al. "The effects of regularizing sleep-wake schedules on daytime sleepiness" (Jun 1996) 19(5) Sleep 432–441—"Subjects in the regular schedule condition reported greater and longer lasting improvements in alertness compared with subjects in the sleep only condition and reported improved sleep efficiency."

Medeiros, A, et al. "The relationships between sleep-wake cycle and academic performance in medical students" (2001) 32(3) Biological Rhythm Research 263–270.

Monash Business School, "How We Behave: Insights from a Trillion Internet Observations" (17 February 2017)

Paterson, JL. "Sleep schedule regularity is associated with sleep duration in older Australian adults: Implications for improving the sleep health and wellbeing of our aging population" (Mar–Apr 2017) 41(2) Clinical Gerontologist 113–122.

Petrov, ME, et al. "Relationship of sleep duration and regularity with dietary intake among preschool-aged children with obesity from low-income families" (Feb–Mar 2017) 38(2) Journal of Developmental and Behavioral Pediatrics 120–128.

Phillips, A, et al. "Irregular sleep/wake patterns are associated with poorer academic performance and delayed circadian and sleep/wake timing" (12 Jun 2017) 7(1) Scientific Reports—"Irregular sleep and light exposure patterns in college students are associated with delayed circadian rhythms and lower academic performance."

RAND Corporation, "Shifting school start times could contribute $83 billion to U.S. economy within a decade" (30 Aug 2017).

Reske, M, et al. "fMRI identifies chronotype-specific brain activation associated with attention to motion—Why we need to know when subjects go to bed" (1 May 2015) 111

NeuroImage 602–610—“Individual sleep preferences are linked to characteristic brain activation patterns.”

Roane, BM, et al. “What role does sleep play in weight gain in the first semester of university?” (12 Aug 2014) 13(56) Behavioral Sleep Medicine 491–505—“Daily variability in sleep duration contributes to males’ weight gain.”

Sano, A, et al. “Influence of sleep regularity on self-reported mental health and wellbeing” (2016) Affective Computing—“Sleep irregularity appears to be associated with lower self-reported mental health and wellbeing (low energy and alertness in the morning), even when controlling for sleep duration and stress.”

Tezler, E, et al. “The effects of poor quality sleep on brain function and risk taking in adolescence” (2013) 71 Neuroimage 275–283.

Vera, B, et al. “Modifiable lifestyle behaviors, but not a genetic risk score, associate with metabolic syndrome in evening chronotypes” (17 Jan 2018) 8(1) Scientific Reports—“Several modifiable factors such as sedentary lifestyle, difficulties in controlling the amount of food eaten, alcohol intake and later wake and bed times that characterized evening-types, may underlie chronotype-MetS relationship.”

Vetter, C, et al. “Prospective study of chronotype and incident depression among middle- and older-aged women in the Nurses’ Health Study II” (25 May 2018) 103 Journal of Psychiatric Research 156–160—“Chronotype may influence the risk of depression in middle- to older-aged women.”

Walch, OJ, et al. “A global quantification of ‘normal’ sleep schedules using smartphone data” (6 May 2016) 2(5) Science Advances—Huge sleep study mapping average wake/sleep patterns around the world.

White, S. “Can night owls become early birds?” (27 Aug 2015) Globe and Mail.

Wong, PM, et al. “Social jetlag, chronotype, and cardiometabolic risk” (Dec 2015) 100(12) Journal of Clinical Endocrinology & Metabolism—“A misalignment of sleep timing is associated with metabolic risk factors that predispose to diabetes and atherosclerotic cardiovascular disease.”

Check phone

Braun Research, “Bank of America Trends in Consumer Mobility Report” (2015)—35 percent said smartphones were the most important thing on their mind when they woke up.

Elhai, JD, et al. “Problematic smartphone use: A conceptual overview and systematic review of relations with anxiety and depression psychopathology” (1 Jan 2017) 207 Journal of Affective Disorders 251–259.

“For most smartphone users, it’s a ‘round-the-clock’ connection” (26 Jan 2017) Industry Today—66 percent of millennials check their phones as soon as they wake.

Gervis, Z. “Going on vacation won’t cure your smartphone addiction” (17 May 2018) New York Post—Report on study that found we check phone 80 times a day on holiday.

Groeger, JA, et al. "Effects of sleep inertia after daytime naps vary with executive load and time of day" (Apr 2011) 125(2) Behavioral Neuroscience 252–260—"Executive functions take longer to return to asymptotic performance after sleep than does performance of simpler tasks which are less reliant on executive functions."

Naftulin, J. "Here's how many times we touch our phones every day" (13 Jul 2016) Business Insider.

"61% people check their phones within 5 minutes after waking up: Deloitte" (29 Dec 2016) BGR.

Brush teeth

Al Makhmari, SA, et al. "Short-term and long-term effectiveness of powered toothbrushes in promoting periodontal health during orthodontic treatment: A systematic review and meta-analysis" (Dec 2017) 152(6) American Journal of Orthodontics and Dentofacial Orthopedics—"No toothbrush type demonstrated clear superiority for gingival health."

Bellis, M. "A Comprehensive History of Dentistry and Dental Care" (19 Mar 2018) ThoughtCo.

Brooks, JK, et al. "Charcoal and charcoal-based dentifrices: A literature review" (7 Jun 2017) 148(9) Journal of the American Dental Association 661–670—No evidence to support charcoal toothpaste.

CADTH, "Community Water Fluoridation Programs: A Health Technology Assessment—Ethical Considerations" (February 2019)—"Overall, this ethics analysis concludes that CWF is ethically justified because it effectively improves public oral health with few harms and side effects."

Cifcibasi, E, et al. "Comparison of manual toothbrushes with different bristle designs in terms of cleaning efficacy and potential role on gingival recession" (2014) 8 Eur J Dent. 395–401—"Bristle design has little impact on plaque removal capacity of a toothbrush."

Government of Canada, "Position statement on Community Water Fluoridation" (23 September 2016)—"Community water fluoridation remains a safe, cost effective and equitable public health practice and an important tool in protecting and maintaining the health and well-being of Canadians."

Knox, MC, et al. "Qualitative investigation of the reasons behind opposition to water fluoridation in regional NSW, Australia" (15 Feb 2017) 27(1) Public Health Research and Practice.

Marinho, VCC, et al. "Fluoride toothpastes for preventing dental caries in children and adolescents" (2003) Cochrane Library.

McLaren, L, et al. "Measuring the short-term impact of fluoridation cessation on dental caries in Grade 2 children using tooth surface indices" (17 Feb 2016) 44(3) Community Dentistry and Oral Epidemiology 274–282.

Mork, N, et al. "Perceived safety and benefit of community water fluoridation: 2009 HealthStyles survey" (2015) 75(4) J Public Health Dent. 327–336—"Twenty-seven percent of respondents reported CWF [community water fluoridation] had no health benefit."

National Health Service, "How to keep your teeth clean" (Nov 25, 2015).

Neelima, M, et al. "'Is powered toothbrush better than manual toothbrush in removing dental plaque?' A crossover randomized double-blind study among differently abled, India" (Mar–Apr 2017) 21(2) Journal of Indian Society of Periodontology 138–143—"Manual toothbrushes were equally effective compared to powered toothbrushes."

O'Mullane, D, et al. "Fluoride and Oral Health" (2016) 33 Community Dental Health 69–99—"Community water fluoridation is safe, effective in caries prevention and very likely to be cost effective . . ."

Perrella, A, et al. "Risk perception, psychological heuristics and the water fluoridation controversy" (2015) 106(4) Canadian Journal of Public Health 197–203—Study that found that opposition to water fluoridation is on the rise.

Ranzan, N, et al. "Are bristle stiffness and bristle end-shape related to adverse effects on soft tissues during toothbrushing? A systematic review" (27 Aug 2018) 69(3) International Dental Journal.

Ritchie, G. "The six month dental recall—Science or legend?" (23 Feb 2018) Science-Based Medicine.

Slade, GD, et al. "Water fluoridation and dental caries in U.S. children and adolescents" (14 Jun 2018) 97(10) Journal of Dental Research 1122–1128—"These findings [which included data from over 7,000 children] confirm a substantial caries-preventive benefit of CWF [community water fluoridation] for U.S. children and that the benefit is most pronounced in primary teeth."

Thornton-Evans, G, et al. "Use of toothpaste and toothbrushing patterns among children and adolescents—United States, 2013–2016" (1 Feb 2019) 68(4) Morbidity and Mortality Weekly Report—"Health care professionals can educate parents about using the recommended amount of fluoride toothpaste under parental supervision to realize maximum benefit."

U.S. Department of Health and Human Services Federal Panel on Community Water Fluoridation, "U.S. Public Health Service Recommendation for Fluoride Concentration in Drinking Water for the Prevention of Dental Caries" (2015) 130 Public Health Rep. 318–331—"Community water fluoridation remains an effective public health strategy for delivering fluoride to prevent tooth decay and is the most feasible and cost-effective strategy for reaching entire communities."

Vibhute, A, et al. "The effectiveness of manual versus powered toothbrushes for plaque removal and gingival health: A meta-analysis" (Apr 2012) 16(2) Journal of Indian Society of Periodontology 156–160—"In general there was no evidence of a statistically significant difference between powered and manual brushes."

Walsh, T, et al. "Fluoride toothpastes of different concentrations for preventing dental caries in children and adolescents" (20 Jan 2010) Cochrane Library.

Yaacob, M, et al. "Powered versus manual toothbrushing for oral health" (17 Jun 2014) Cochrane Library—"Powered toothbrushes reduce plaque and gingivitis more than manual toothbrushing in the short and long term. The clinical importance of these findings remains unclear."

Check phone, again

"Americans check their phones 80 times a day: Study" (8 Nov 2017) New York Post.

Bhandari, K. "College students in India check their phones over 150 times daily" (22 May 2018) Toronto Star.

Brandon, J. "The surprising reason millennials check their phones 150 times a day" (17 Apr 2017) Inc.—There are numerous studies on how much we check our phones. I looked at many and put the average at around 100.

"Mobile users cannot leave their phone alone for six minutes, check it 150 times a day" (11 Feb 2013) News.

Ulger, F, et al. "Are we aware how contaminated our mobile phones with nosocomial pathogens?" (2009) 7 Annals of Clinical Microbiology and Antimicrobials—"In total, 94.5% of phones demonstrated evidence of bacterial contamination with different types of bacteria."

Step on the bathroom scale

American Heart Association, "Daily weighing may be key to losing weight" (5 Nov 2018)—Study summary.

Amy, NK, et al. "Barriers to routine gynecological cancer screening for White and African-American obese women" (Jan 2006) 30(1) International Journal of Obesity 147–155—Study notes that women with obesity might avoid care because of shame associated with weighing.

Aydinoğlu, NZ, et al. "Imagining thin: Why vanity sizing works" (Oct 2012) 22(4) Journal of Consumer Psychology 565–572—"We find that fitting into a pair of jeans labeled smaller than its true size can increase positive self-related mental imagery for consumers."

Benn, Y, et al. "What is the psychological impact of self-weighing? A meta-analysis" (9 Feb 2016) 10(2) Health Psychology Review 187–203.

Bigotti, F. "Mathematica medica: Santorio and the quest for certainty in medicine" (22 Jul 2016) 1(4) Journal of Healthcare Communications.

Bivins, R, et al. "Weighting for Health: Management, measurement and self-surveillance in the modern household" (Nov 2016) 29(4) Social History of Medicine 757–780—Terrific history of the bathroom scale. Quote on early advertisement: "It's a national duty to keep fit. Check your weight daily."

Boo, S. "Misperception of body weight and associated factors" (2014) 16(4) Nursing & Health Sciences 468–475—"48.9% underestimating and 6.8% overestimating their weight status."

Boseley, S. "Third of overweight teenagers think they are right size, study shows" (9 Jul 2015) Guardian.

Burke, MA, et al. "Evolving societal norms of obesity: What is the appropriate response?" (16 Jan 2018) 319(3) JAMA 221–222.

Butryn, ML, et al. "Consistent self-monitoring of weight: a key component of successful weight loss maintenance" (18 Jan 2007) Obesity 15(12)—"Consistent self-weighing may help individuals maintain their successful weight loss."

Chrisafis, A. "French women Europe's thinnest and most worried about weight, report finds" (23 Apr 2009) Guardian—45 percent unhappy with weight.

Dahl, M. "Six-pack stress: Men worry more about their appearance than their jobs" (28 Feb 2014) Today—63 percent of men said they "always feel like (they) could lose weight."

Ethan, D, et al. "An analysis of weight loss articles and advertisements in mainstream women's health and fitness magazines" (2016) 6(2) Health Promotion Perspectives 80–84—"Themes commonly noted in the advertisements for weight loss products appealed to appearance-based motivations that may have deleterious effects on women's weight loss perceptions and behaviors."

"Fury at a Canadian university after it removes scales from the campus gym because they are 'triggering eating disorders'" (15 Mar 2017) Daily Mail.

Greaves, C, et al. "Understanding the challenge of weight loss maintenance: A systematic review and synthesis of qualitative research on weight loss maintenance" (7 Apr 2017) 11(2) Health Psychology Review 145–163.

Howe, C, et al. "Parents' Underestimations of Child Weight: Implications for Obesity Prevention" (2017) Journal of Pediatric Nursing 57–61—"96% of parents underestimated their overweight children."

Hurst, M, et al. "'I just feel so guilty': The role of introjected regulation in linking appearance goals for exercise with women's body image" (Mar 2017) 20 Body Image 120–129—"Appearance goals for exercise are consistently associated with negative body image."

Ingraham, C. "The absurdity of women's clothing sizes, in one chart" (11 Aug 2015) Washington Post—"A size 8 dress today is nearly the equivalent of a size 16 dress in 1958."

Ingraham, C. "Nearly half of America's overweight people don't realize they're overweight" (1 Dec 2016) Washington Post.

Jackson, SE. "Weight perceptions in a population sample of English adolescents: Cause for celebration or concern?" (Oct 2015) 39(10) International Journal of Obesity 1488–1493.

Jospe, MR, et al. "The effect of different types of monitoring strategies on weight loss: A randomized controlled trial" (Sep 2017) 25(9) Obesity 1490–1498.

Katterman, SN, et al. "Daily weight monitoring as a method of weight gain prevention in healthy weight and overweight young adult women" (11 Jun 2015) 21(12) Journal of Health Psychology—"No harmful effects of daily weighing were detected; acceptability and adherence were high. Weight monitoring did not impact weight; both groups showed little weight gain. Results suggest that weight monitoring has minimal harmful effects and may be useful for preventing weight gain."

Ketron, S. "Consumer cynicism and perceived deception in vanity sizing: The moderating role of retailer (dis)honesty" (Nov 2016) 33(C) Journal of Retailing and Consumer Services 33–42.

Ketron, S, et al. "Liar, liar, my size is higher: How retailer context influences labeled size believability and consumer responses to vanity sizing" (Jan 2017) 34(C) Journal of Retailing and Consumer Services 185–192.

Lanoye, A, et al. "Young adults' attitudes and perceptions of obesity and weight management: Implications for treatment development" (Mar 2016) 5(1) Current Obesity Reports 14–22—"Weight management interventions targeting young adults should continue to include screening for previous disordered eating practices and excessive body dissatisfaction, but frequent self-weighing within the context of supervised weight gain prevention or weight loss trials appears to be appropriate and beneficial for most young adults."

LaRose, JG, et al. "Frequency of self-weighing and weight loss outcomes within a brief lifestyle intervention targeting emerging adults" (Mar 2016) 2(1) Obesity Science & Practice 88–92—"Consistent with findings among other adult samples, frequent self-weighing was associated with greater weight losses."

Madigan, CD, et al. "A randomised controlled trial of the effectiveness of self-weighing as a weight loss intervention" (10 Oct 2014) 11 International Journal of Behavioural Nutrition and Physical Activity—"As an intervention for weight loss, instruction to weigh daily is ineffective. Unlike other studies, there was no evidence that greater frequency of self-weighing is associated with greater weight loss."

Madigan, CD, et al. "Regular self-weighing to promote weight maintenance after intentional weight loss: A quasi-randomized controlled trial" (Jun 2014) 36(2) Journal of Public Health 259–267—"Encouraging people who have recently lost weight to weigh themselves regularly prevents some weight regain."

Mercurio, A, et al. "Watching my weight: Self-weighing, body surveillance, and body dissatisfaction" (Jul 2011) 65 (1-2) Sex Roles 47–55—Self-weighing may be psychologically more harmful to young women,"a group that is often striving to meet thinness ideals and current beauty standards."

Pacanowski, CR, et al. "Self-weighing: Helpful or harmful for psychological well-being? A review of the literature" (Mar 2015) 4(1) Current Obesity Reports 65–72.

Pacanowski, CR, et al. "Self-weighing throughout adolescence and young adulthood:

Implications for well-being” (Nov–Dec 2015) 47(6) Journal of Nutrition Education and Behavior—“Self-weighing may not be an innocuous behavior for young people, particularly women. Interventions should assess potential harmful consequences of self-weighing in addition to any potential benefits.”

Putterman, E, et al. “Appearance versus health: Does the reason for dieting affect dieting behavior?” (Apr 2004) 27(2) Journal of Behavioral Medicine 185–204.

Reeves, S. “Santorio Santorio—physician, physiologist, and weight-watcher” (2016) 8(1) Hektoen International.

Robinson, E, et al. “Perceived weight status and risk of weight gain across life in US and UK adults” (1 Dec 2015) 39(12) International Journal of Obesity 1721–1726—“Perceiving oneself as being ‘overweight’ is counter-intuitively associated with an increased risk of future weight gain among US and UK adults.”

Rosenbaum, DL, et al. “Daily self-weighing and weight gain prevention: A longitudinal study of college-aged women” (Oct 2017) 40(5) Journal of Behavioural Medicine 846–853.

Shieh, C. “Self-weighing in weight management interventions: A systematic review of literature” (Sep–Oct 2016) 10(5) Obesity Research & Clinical Practice—“Self-weighing is likely to improve weight outcomes, particularly when performed daily or weekly, without causing untoward adverse effects. Weight management interventions could consider including this strategy.”

Snook, K, et al. “Change in percentages of adults with overweight or obesity trying to lose weight, 1988–2014” (7 Mar 2017) 317(9) Journal of the Association of American Medicine—Fewer adults trying to lose weight.

Steinberg, DM, et al. “Daily self-weighing and adverse psychological outcomes: A randomized controlled trial” (Jan 2014) 46 American Journal of Preventive Medicine—“Self-weighing is not associated with adverse psychological outcomes.”

Steinberg, DM, et al. “Weighing everyday matters: Daily weighing improves weight loss and adoption of weight control behaviors” (Apr 2015) 115(4) Journal of the Academy of Nutrition and Dietetics 511–518—“Weighing everyday led to greater adoption of weight control behaviors and produced greater weight loss compared to weighing most days of the week. This further indicates daily weighing as an effective weight loss tool.”

Stewart, TM. “Why thinking we’re fat won’t help us improve our health: Finding the middle ground” (Jul 2010) 26(7) Obesity [illegible]—Terrific commentary outlining tension between weight bias, normalized larger weights, and weight loss.

Wilke, J. “Nearly half in U.S. remain worried about their weight” (25 Jul 2014) Gallup.com—“Almost half of Americans (45%) worry about their weight ‘all’ or ‘some of the time,’ significantly higher than the 34% who reported this level of worry in 1990.”

Wilkinson, L, et al. “Three-year follow-up of participants from a self-weighing randomized controlled trial” (19 Sep 2017) Journal of Obesity—“Frequent self-weighing may be an effective, low-cost strategy for weight loss maintenance.”

Williams, N. "Trying to lose weight? Then ditch the scale, says P.E.I. dietitian" (15 Jan 2018) CBC News.

Wing, RR, et al. "Frequent self-weighing as part of a constellation of healthy weight control practices in young adults" (May 2015) 23(5) Obesity 943–949—"Frequent weighing was associated with healthy weight management strategies, but not with unhealthy practices or depressive symptoms."

Zheng, Y, et al. "Self-weighing in weight management: A systematic literature review" (Feb 2015) 23(2) Obesity 256–265—"Regular self-weighing has been associated with weight loss and not with negative psychological outcomes."

Get dressed

"Briefs are the underwear of choice among both men and women" (14 Dec 2017) YouGov.

Feldman, J. "More people go commando than you might think" (10 Mar 2014) HuffPost—"7 percent said they went without undergarments 'all the time.'"

Gunter, J. "Is it important to wear 100 percent cotton underwear for optimal vaginal health, or does it not matter?" (2019) New York Times.

Hamlin, AA, et al. "Brief versus Thong Hygiene in Obstetrics and Gynecology (B-THONG): A survey study" (Jun 2019) 45(6) Journal of Obstetrics and Gynaecology Research—"Oral sex was the only independent predictor of urinary tract infection and bacterial vaginosis."

Mínguez-Alarcón, L, et al. "Type of underwear worn and markers of testicular function among men attending a fertility center" (1 Sep 2018) 33(9) Human Reproduction—"Certain styles of male underwear may impair spermatogenesis."

Sapra, KJ, et al. "Choice of underwear and male fecundity in a preconception cohort of couples" (May 2016) 4(3) Andrology—"No significant differences in time-to-pregnancy, conception delay, or infertility were observed. In summary, male underwear choice is associated with few differences in semen parameters; no association with time-to-pregnancy is observed."

Sapra, KJ, et al. "Male underwear and semen quality in a population-based preconception cohort" (20 Oct 2015) 104(3) Fertility and Sterility—"Better semen quality parameters are observed in men wearing boxers during the day and none to bed."

Coffee

Stromberg, J. "It's a myth: There's no evidence that coffee stunts kids' growth" (20 Dec 2013) Smithsonian.

World Health Organization International Agency for Research on Cancer, "IARC Monographs evaluate drinking coffee, maté, and very hot beverages" (15 Jun 2016)—"After thoroughly reviewing more than 1000 studies in humans and animals, the Working Group found that there was *inadequate evidence* for the carcinogenicity of coffee drinking overall."

Breakfast

Adolphus, K, et al. "The effects of breakfast on behavior and academic performance in children and adolescents" (8 Aug 2013) 7 Frontiers in Human Neuroscience 425—"Some evidence suggested that quality of habitual breakfast, in terms of providing a greater variety of food groups and adequate energy, was positively related to school performance."

Allen, V. "Breakfast IS key to losing weight" (23 Apr 2018) Daily Mail—Inaccurate headline associated with a small, unpublished study.

Barr, SI, et al. "Association of breakfast consumption with body mass index and prevalence of overweight/obesity in a nationally-representative survey of Canadian adults" (31 Mar 2016) 15 Nutrition Journal—"Among Canadian adults, breakfast consumption was not consistently associated with differences in BMI or overweight/obesity prevalence."

Bohan Brown, MM, et al. "Eating compared to skipping breakfast has no discernible benefit for obesity-related anthropometrics: Systematic review and meta-analysis of randomized controlled trials" (1 Apr 2017) 31(1) Federation of American Societies for Experimental Biology Journal.

Brown, AW, et al. "Belief beyond the evidence: Using the proposed effect of breakfast on obesity to show 2 practices that distort scientific evidence" (Nov 2013) 98(5) American Journal of Clinical Nutrition 1298–1308—"The belief in the PEBO [proposed effect of breakfast on obesity] exceeds the strength of scientific evidence."

Carroll, AE. "Sorry, there's nothing magical about breakfast" (23 May 2016) New York Times—Nice review of the literature.

Cheng, E, et al. "Offering breakfast in the classroom and children's weight outcomes" (25 Feb 2019) 173(4) JAMA Pediatrics.

Chowdhury, E, et al. "Six weeks of morning fasting causes little adaptation of metabolic or appetite resources to feeding in adults with obesity" (May 2019) 27(5) Obesity—"There was little evidence of metabolic adaptation to acute feeding or negative consequences from sustained morning fasting. This indicates that previously observed differences between breakfast consumers and skippers may be acute effects of feeding or may have resulted from other lifestyle factors."

Dhurandhar, EJ, et al. "The effectiveness of breakfast recommendations on weight loss: A randomized controlled trial" (2014) 100(2) American Journal of Clinical Nutrition 507–513—Study found no "discernable effect on weight loss."

Levitsky, DA, et al. "Effect of skipping breakfast on subsequent energy intake" (2 Jul 2013) 119 Physiology & Behavior 9–16—"Skipping breakfast was not compensated by an increase in intake at lunch. Consequently, total daily energy intake was reduced by skipping breakfast."

Polonsky, HM, et al. "Effect of a breakfast in the classroom initiative on obesity in urban school-aged children: A cluster randomized clinical trial" (1 Apr 2019) 173(4) JAMA Pediatrics—"The initiative had an unintended consequence of increasing incident and prevalent obesity."

Rong, S, et al. "Association of skipping breakfast with cardiovascular and all-cause mortality" (30 Apr 2019) 73(16) Journal of the American College of Cardiology—"Our study supports the benefits of eating breakfast in promoting cardiovascular health."

Shimizu, H, et al. "Delayed first active-phase meal, a breakfast-skipping model, led to increased body weight and shifted the circadian oscillation of the hepatic clock and lipid metabolism-related genes in rats fed a high-fat diet" (31 Oct 2018) 13(10) PloS ONE.

Sievert, K, et al. "Effect of breakfast on weight and energy intake: Systematic review and meta-analysis of randomised controlled trials" (30 Jan 2019) 364 BMJ—"The addition of breakfast might not be a good strategy for weight loss, regardless of established breakfast habit. Caution is needed when recommending breakfast for weight loss in adults, as it could have the opposite effect."

"Skipping breakfast makes you fat" (26 Apr 2018) Times Now.

Spector, T. "Breakfast—the most important meal of the day?" (30 Jan 2019) BMJ Opinion—"The disadvantages of skipping breakfast have now been debunked by several randomised trials."

St-Onge, MP, et al. "Meal timing and frequency: Implications for cardiovascular disease prevention: A scientific statement from the American Heart Association" (28 Feb 2017) 135(9) Circulation—Concludes that eating breakfast is correlated with a range of health benefits.

Milk

Barker, ME, et al. "What type of milk is best? The answer is to follow your tastebuds" (2 Jan 2018) Independent—Excellent review that concludes, "Whole milk has more calories but there's little evidence skimmed is healthier."

Centers for Disease Control and Prevention, "Increased outbreaks associated with non-pasteurized milk, 2007–2012" (8 Jun 2017).

Centers for Disease Control and Prevention, "Raw milk: Know the raw facts."

Claeys, WL, et al. "Raw or heated cow milk consumption: Review of risks and benefits" (May 2013) 31(1) Food Control 251–262—"Consumption of raw milk poses a realistic health threat due to a possible contamination with human pathogens. It is therefore strongly recommended that milk should be heated before consumption . . . Heating (in particularly ultra high temperature and similar treatments) will not substantially change the nutritional value of raw milk or other benefits associated with raw milk consumption."

Collier, R. "Dairy research: 'Real' science or marketing?" (12 Jul 2016) 188(10) Canadian Medical Association Journal 715–716.

De Oliveira Otto, MC, et al. "Serial measures of circulating biomarkers of dairy fat and total and cause-specific mortality in older adults: The Cardiovascular Health Study" (Sep 2018) 108(3) American Journal of Clinical Nutrition.

Dehghan, M, et al. "Association of dairy intake with cardiovascular disease and mortality in 21 countries from five continents (PURE): A prospective cohort study" (11 Sep 2018) 292(10161) The Lancet—"Dairy consumption was associated with lower risk of mortality and major cardiovascular disease events in a diverse multinational cohort."

Drouin-Chartier, JP, et al. "Comprehensive review of the impact of dairy foods and dairy fat on cardiometabolic risk" (15 Nov 2016) 7(6) Advances in Nutrition 1041–1051.

Fenton, TR, et al. "Milk and acid-base balance: Proposed hypothesis versus scientific evidence" (Oct 2011) 30(5 Suppl 1) Journal of the American College of Nutrition, 471S–475S—"The modern diet, and dairy product consumption, does not make the body acidic."

Food and Drug Administration, "The dangers of raw milk: Unpasteurized milk can pose a serious health risk" (11 Aug 2018).

Food and Drug Administration, "Milk residue sampling survey" (Mar 2015).

Hamblin, J. "How agriculture controls nutrition guidelines" (8 Oct 2015) Atlantic.

Holmberg, S, et al. "High dairy fat intake related to less central obesity: A male cohort study with 12 years' follow-up" (Jun 2013) 31(2) Scandinavian Journal of Primary Health Care 89–94—"A high intake of dairy fat was associated with a lower risk of central obesity and a low dairy fat intake was associated with a higher risk of central obesity."

Jalonick, MC. "Little evidence of antibiotics in U.S. milk supply: FDA" (5 Mar 2015) CTV News.

Lucey, J. "Raw milk consumption: Risks and benefits" (Jul–Aug 2015) 50(4) Nutrition Today 189–193—Useful review of evidence. "Claims related to improved nutrition, prevention of lactose intolerance, or provision of 'good' bacteria from the consumption of raw milk have no scientific basis and are myths."

Markham, L, et al. "Attitudes and beliefs of raw milk consumers in northern Colorado" (24 Nov 2014) 9(4) Journal of Hunger & Environmental Nutrition 546–564.

Matthews-King, A. "Belief that milk makes cold mucus and phlegm worse is a medieval myth, scientists say" (7 Sep 2018) Independent.

Michaëlsson, K, et al. "Milk intake and risk of mortality and fractures in women and men: Cohort studies" (28 Oct 2014) 349 BMJ.

Mole, B. "Raw milk is trending for some reason—so are nasty, drug-resistant infections" (11 Feb 2018) Ars Technica.

"More families say no to cow's milk" (2 Feb 2015) CBS News—"The average consumption of dairy milk has dropped from about 22 gallons a year per person in 1970 to less than 15 gallons in 2012. That's a 33 percent decline."

Mullie, P, et al. "Daily milk consumption and all-cause mortality, coronary heart disease and stroke: A systematic review and meta-analysis of observational cohort studies" (8 Dec 2016) 16(1) BMC Public Health—"No evidence for a decreased or increased risk of all-cause mortality, coronary heart disease, and stroke associated with adult milk consumption. However, the possibility cannot be dismissed that risks associated with milk consumption could be underestimated because of publication bias."

National Health Service, "Dairy and alternatives in your diet" (16 Jan 2018).

New Zealand, Office of the Prime Minister's Chief Science Advisor, "Review of evidence for health benefits of raw milk consumption" (May 2015).

Rahn, W, et al. "Framing food policy: The case of raw milk" (31 Mar 2016) 45(2) Policy Studies Journal—"Consumer choice and the virtues of small farms (e.g., 'locavorism' in the words of the food movement) are powerful frames for raw milk advocates; the claims of public health authorities and representatives of the traditional dairy industry, conversely, do not fare well when people have access to alternative ways to understand the issue."

Rangwani, S. "White poison: The horrors of milk" (3 December 2001).

Rautiainen, S, et al. "Dairy consumption in association with weight change and risk of becoming overweight or obese in middle-aged and older women: A prospective cohort study" (Apr 2016) 103(4) American Journal of Clinical Nutrition 979–988.

Richardson, SB. "Legal pluralism and the regulation of raw milk sales in Canada: Creating space for multiple normative orders at the food policy table" in Alabrese, M, et al. (eds), *Agricultural Law: Current Issues from a Global Perspective* (Oct 2017).

Rozenberg, S, et al. "Effects of dairy products consumption on health: Benefits and beliefs—A commentary from the Belgian Bone Club and the European Society for Clinical and Economic Aspects of Osteoporosis, Osteoarthritis and Musculoskeletal Diseases" (Jan 2016) 98(1) Calcified Tissue International 1–17.

Saini, V, et al. "Antimicrobial use on Canadian dairy farms" (Mar 2012) 95(3) Journal of Dairy Science.

St. Pierre, M. "Changes in Canadians' preferences for milk and dairy products" (21 Apr 2017) Statistics Canada.

Thorning, TK, et al. "Milk and dairy products: Good or bad for human health? An assessment of the totality of scientific evidence" (22 Nov 2016) 60 Food & Nutrition Research—"Intake of milk and dairy products contribute to meet nutrient recommendations, and may protect against the most prevalent chronic diseases, whereas very few adverse effects have been reported."

Yakoob, MY, et al. "Circulating biomarkers of dairy fat and risk of incident diabetes mellitus among US men and women in two large prospective cohorts" (26 Apr 2016) 133(17) Circulation—"In 2 prospective cohorts, higher plasma dairy fatty acid concentrations were associated with lower incident diabetes."

Zylberberg, D. "Milk, ideology, and law: Perfect foods and imperfect regulation" (2016) 104 Georgetown Law Journal 1377.

Kids off to school

Bennett, D. "Abducted!" (20 Jul 2008) Boston Globe—"The Amber Alert system is more effective as theater than as a way to protect children."

Brody, JE. "Parenting advice from 'America's Worst Mom'" (15 Jan 2015) New York Times.

Brussoni, M, et al. "What is the relationship between risky outdoor play and health in children? A systematic review" (8 Jun 2015) 12(6) International Journal of Environmental Research and Public Health 6423–6454—Study supporting risky outdoor play.

Cairns, W. *How to Live Dangerously* (2008) Macmillan—Contains stat about leaving child outside for 200,000 years.

Centers for Disease Control and Prevention, National Center for Injury Prevention and Control, "10 leading causes of injury deaths by age group" (2015).

"Children who travel to school independently are more satisfied and perform better in school" (19 Jan 2018) Medical XPress—Children who travel with their parents' supervision "lose natural opportunities to explore their neighborhood and to interact with friends on their own. As a result, they become less independent and secure in their immediate environment."

De La Cruz, D. "Utah passes 'free-range' parenting law" (29 Mar 2018) New York Times.

DeSilver, D. "Dangers that teens and kids face: A look at the data" (14 Jan 2016) Pew Research Center.

Donovan, K, et al. "The role of entertainment media in perceptions of police use of force" (17 Sep 2015) 42 Criminal Justice and Behaviour—Study notes influence of crime TV on perceptions of crime and criminal justice.

Eagle Shutt, J, et al. "Reconsidering the leading myths of stranger child abduction" (26 Jan 2004) 17 Criminal Justice Studies 127–134.

Faulkner, GEJ, et al. "What's 'quickest and easiest?': Parental decision making about school trip mode" (6 Aug 2010) 7 International Journal of Behavioral Nutrition and Physical Activity—Interview study that produced quotes on concerns associated with strangers.

Foster, S, et al. "Suspicious minds: Can features of the local neighbourhood ease parents' fears about stranger danger?" (18 Feb 2015) 42 Journal of Environmental Psychology 48–56—Parents' fears about strangers restrict child mobility, and more walkable neighborhoods would both create safer streets and reduce parental fear. Study also found that many parents recognize that their fear does not fit actual risks.

Fridberg, L. "Children who travel to school independently are more satisfied and perform better in school" (19 Jan 2018) Karlstad University Service Research Center.

Gainforth, HL, et al. "Evaluating the ParticipACTION 'Think Again' Campaign" (Aug 2016) 43 Health Education & Behavior [illegible]—Notes that [illegible] percent of parents think kids get enough exercise, when only 7 percent do.

García-Hermoso, A, et al. "Associations between the duration of active commuting to school and academic achievement in rural Chilean adolescents" (2017) 22(1) Environmental Health and Preventive Medicine—Active commuting "may have a positive influence on academic achievement in adolescents."

Griffin, T, et al. "Does AMBER Alert 'save lives'? An empirical analysis and critical implications" (4 Feb 2015) 39(4) Journal of Crime and Justice—AMBER alerts had no direct impact on outcome, though likely helped in approximately 25 percent of cases, typically cases involving an abduction by a family member.

Herrador-Colmenero, M, et al. "Children who commute to school unaccompanied have greater autonomy and perceptions of safety" (Dec 2017) 106(2) Acta Pædiatrica—Active commuting to school (alone) resulted in more safety awareness.

Huertas-Delgado, FJ, et al. "Parental perceptions of barriers to active commuting to school in Spanish children and adolescents" (27 Sep 2017) 106(12) European Journal of Public Health 416–421.

Iancovich, V. "Why walking to school is better than driving for your kids" (11 Sep 2015) U of T News—Quotes researcher George Mammen: "Evidence shows that children are more likely to be harmed in a car accident compared to walking to school."

Ingraham, C. "There's never been a safer time to be a kid in America" (14 Apr 2015) Washington Post.

Jacobs, T. "AMBER Alerts largely ineffective, study shows" (15 Dec 2007) Pacific Standard.

Jamieson, PE, et al. "Violence in popular U.S. prime time TV dramas and the cultivation of fear: A time series analysis" (17 Jun 2014) 2(2) Media and Communication 31–41— "Annual change in TV violence, after controlling for the violent crime rate and perceptions of crime prevalence, was significantly related to change in national fear of crime from 1972 to 2010."

Jessup, C, et al. "Fear, hype, and stereotypes: Dangers of overselling the Amber Alert program" (5 Jan 2015) 8 Albany Government Law Review 467–507.

Kort-Butler, LA, et al. "Watching the detectives: Crime programming, fear of crime, and attitudes about the criminal justice system" (2011) 52 (1) Sociological Quarterly 36–55—Study highlights the complex relationship between crime TV and public attitudes about crime.

Lambiase, MJ, et al. "Effect of a simulated active commute to school on cardiovascular stress reactivity" (Aug 2010) 42(8) Medicine & Science in Sports & Exercise 1609–1616—Walking to school may help with stress and cardiovascular health.

Luymes, G. "School zone driving is getting worse in B.C., survey suggests" (5 Sep 2017) Vancouver Sun—Report on survey exploring problems with school drop-offs, including speeding, failing to stop, etc.

Martínez-Gómez, D, et al. "Active commuting to school and cognitive performance in adolescents" (Apr 2011) Archives of Pediatrics and Adolescent Medicine 300–305—"These novel results contribute to a growing body of evidence indicating that physical activity may have a beneficial influence on cognition in youth."

McDonald, NC, et al. "Why parents drive children to school: Implications for safe routes to school programs" (30 Jun 2009) 75(3) Journal of the American Planning Association 331–342—Fear of strangers is a dominant reason parents drive kids to school.

Miller, MK, et al. "The psychology of AMBER Alert: Unresolved issues and implications" (Mar 2009) 46(1) The Social Science Journal 111–123.

Moore, A. "Walking, biking to school improves concentration" (24 Nov 2012) Medical

Daily—Reports on a study of almost 20,000 students. Found an improvement in concentration.

Moscowitz, L, et al. "'Every parent's worst nightmare': Myths of child abductions in US news" (2011) 5(2) Journal of Children and Media—Study found a bias in media reporting of child abductions: "Caucasian girls being snatched from their middle- to upper-class homes by male strangers."

Neufeld, L. "School zone speed limits reduce collisions and injuries, says city" (13 Oct 2016) CBC News—Story notes that there were two pedestrian school-zone injuries in 2016.

Press Association, "Parents face fines for driving children to school in push to curb pollution" (6 Sep 2017) Guardian.

Rothman, L, et al. "Dangerous student car drop-off behaviors and child pedestrian–motor vehicle collisions: An observational study" (3 Jul 2016) 17(5) Traffic Injury Prevention 454–459—"Dangerous drop-off behaviors were observed in 104 schools (88%)." Study also notes, "In Toronto, Canada, there are approximately 150 PMVCs [pedestrian motor vehicle collisions] per year in school-age children."

Rothman, L, et al. "Motor vehicle-pedestrian collisions and walking to school: The role of the built environment" (May 2014) 133(5) Pediatrics—Study found walking to school to be safe and that physicians should "counsel parents to encourage children to walk to school as a healthy lifestyle choice." Study also found 481 collisions in Toronto area, most minor. There were 30 major injuries and 1 fatality.

Shalev, GK, et al. "The more eyes the better? A preliminary examination of the usefulness of child alert systems in the Netherlands, United Kingdom (UK), Czech Republic and Poland" (Mar 2016) University of Portsmouth, Centre for the Study of Missing Persons.

Sicafuse, LL, et al. "Social psychological influences on the popularity of Amber Alerts" (30 Sep 2010) 37(11) Criminal Justice and Behavior 1237–1254—"Because AMBER Alerts are illusory means of controlling crime, they may be conceptualized as 'crime control theater' (CCT) and thus are indicative of a problematic social tendency to address complex issues through simple solutions."

"Single dad barred from sending kids to school on city bus" (6 Sep 2017) CTV News.

Skenazy, L. "Why I let my 9-year-old ride the subway alone" (1 Apr 2008) New York Sun.

Smith, LJ. "Parents could land up to £1,000 in fines for driving their kids to school, this is why" ([illegible] May 2017) Express.

Stokes, MA. "Stranger danger: Child protection and parental fears in the risk society" (2009) Amsterdam Social Science 1(3) 6–24—Article analyzes stranger danger and views it as the "modern embodiment of the 'wicked witch' or 'bogeyman.'"

Teschke, K, et al. "Risks of cycling, walking and driving put in context" (7 Aug 2013) Vancouver Sun.

Thomas, AJ, et al. "Correction: No child left alone: Moral judgments about parents affect estimates of risk to children" (23 Aug 2016) 2(1) Collabra 1–15.

Tremblay, MS, et al. "Position statement on active outdoor play" (8 Jun 2015) 12(6) International Journal of Environmental Research and Public Health 6475–6505—Consensus statement that estimated the risk of abduction at 1 in 14 million.

Valentine, G. "'Oh yes I can.' 'Oh no you can't': Children and parents' understandings of kids' competence to negotiate public space safely" (Jan 1997) 29(1) Antipode.

Westman, J, et al. "What drives them to drive?—Parents' reasons for choosing the car to take their children to school" (8 Nov 2017) 8 Frontiers in Psychology, https://doi.org/10.3389/fpsyg.2017.01970—Study notes that distance to school not a big factor in decision to drive. Also found fear of strangers to be a dominant concern.

Getting to work

Andersen, LB. "Active commuting is beneficial for health" (19 Apr 2017) 357 BMJ—Review of benefits of commuting, including 30 percent reduction in all-cause mortality.

Barth, B. "Why biking to work is a barrier for most Americans" (6 Apr 2015) Momentum Mag.

Breakaway Research Group, "U.S. bicycling participation benchmarking study report" (Mar 2015)—Article reviews survey of Americans on barriers to cycling.

Bubbers, M. "What bicycle-friendly Copenhagen can teach us about commuting" (7 Mar 2018) Globe and Mail.

Celis-Morales, CA, et al. "Association between active commuting and incident cardiovascular disease, cancer, and mortality: Prospective cohort study" (19 Apr 2017) 357 BMJ—"Cycle commuting was associated with a lower risk of CVD, cancer, and all cause mortality. Walking commuting was associated with a lower risk of CVD independent of major measured confounding factors."

Cheung, A. "Drivers vs. cyclists—who's at fault? New study reveals who Canadians blame for road dangers" (28 Jun 2018) CBC News—"Data from a 2015 report from London, England's Department of Transport suggested that 60 to 70 per cent of the time drivers were solely responsible, while cyclists were to blame 17 to 20 per cent of the time."

De Hartog, JJ, et al. "Do the health benefits of cycling outweigh the risks?" (Aug 2010) 118(8) Environmental Health Perspectives 1109–1116—"The estimated health benefits of cycling were substantially larger than the risks."

Department for Transport, "British Social Attitudes Survey 2014" (2014)—"In 2014, 64% of respondents agreed that it is too dangerous for them to cycle on the road."

Dinu, M, et al. "Active commuting and multiple health outcomes: A systematic review and meta-analysis" (Mar 2019) 49(3) Sports Medicine—"People who engaged in active commuting had a significantly reduced risk of all-cause mortality, cardiovascular disease incidence and diabetes."

Fruhen, LS, et al. "Car driver attitudes, perceptions of social norms and aggressive driving behaviour towards cyclists" (Oct 2015) 83 Accident Analysis & Prevention 162–170.

Goddard, T. "Exploring drivers' attitudes and behaviors toward bicyclists: The effect of explicit and implicit attitudes on self-reported safety behaviors" (Dec 2015) Portland State University Transportation and Research Center.

Johnson, M, et al. "Riding through red lights: The rate, characteristics and risk factors of non-compliant urban commuter cyclists" (Jan 2011) 43(1) Accident Analysis & Prevention 323–328—"A cross-sectional observational study was conducted using a covert video camera to record cyclists at 10 sites across metropolitan Melbourne, Australia from October 2008 to April 2009. In total, 4225 cyclists faced a red light and 6.9% were non-compliant."

Kahneman, D, et al. "A survey method for characterizing daily life experience: The day reconstruction method." (3 Dec 2004) 306(5702) Science 1776–1780.

Kemp, M. "Four in every five crashes between cars and bicycles caused by driver of car" (19 Feb 2013) AdelaideNow.

Lin, P, et al. "Naturalistic bicycling behavior pilot study" (Nov 2017) University of South Florida Center for Urban Transportation Research—"The proportion of compliance with general traffic rules for bicyclists was 88.1% in the daytime and 87.5% in the nighttime."

Macmillan, A, et al. "Trends in local newspaper reporting of London cyclist fatalities 1992–2012: The role of the media in shaping the systems dynamics of cycling" (Jan 2016) 86 Accident Analysis & Prevention 137–145—Stats on coverage of cycling accidents.

Manning, J. "A short commute makes Londoners happier than sex" (30 Jan 2018) Time Out.

Marshall, W, et al. "Scofflaw bicycling: Illegal but rational" (Apr 2017) 11(1) Journal of Transport and Land Use, 805–836—Study finds that "bicyclists similarly rationalize their illegal behaviors but were more inclined to cite increasing their own personal safety and/or saving energy."

Martin, A, et al. "Does active commuting improve psychological wellbeing? Longitudinal evidence from eighteen waves of the British Household Panel Survey" (Dec 2014) 69 Preventive Medicine 296–303.

Morency, P, et al. "Traveling by bus instead of car on urban major roads: Safety benefits for vehicle occupants, pedestrians, and cyclists" (Apr 2018) 95(2) Journal of Urban Health 196–207—"City bus is a safer mode than car, for vehicle occupants but also for cyclists and pedestrians traveling along these bus routes."

Mueller, N, et al. "Health impact assessment of active transportation: A systematic review" (Jul 2015) 76 Preventive Medicine 103–114—"Physical activity benefits exceed traffic-associated collision/emission detriments."

Mullan, E. "Exercise, weather, safety, and public attitudes: A qualitative exploration of leisure cyclists' views on cycling for transport" (16 Jul 2013) Sage Open—"The general public had very negative attitudes to cycling and cyclists."

Panter, J, et al. "Using alternatives to the car and risk of all-cause, cardiovascular and cancer mortality" (2018) Heart—"More active patterns of travel were associated with a [30%] reduced risk of incident and fatal CVD and all-cause mortality in adults."

Puentes, R. "How commuting is changing" (18 Sep 2017) US News—76.3 percent of Americans commute alone in a car.

Seyda, L, et al. "Spoke'n word: A qualitative exploration of the image and psychological factors that deter bicycle commuting" (11 Jan 2015)—Paper delivered at the 94th Annual Meeting of the Transportation Research Board.

Sims, D, et al. "Predicting discordance between perceived and estimated walk and bike times among university faculty, staff, and students" (10 Jan 2018) 14(8) Transportmetrica A: Transport Science—Study finding that 93 percent of workers misestimated biking times to various locations.

St-Louis, E, et al. "The happy commuter: A comparison of commuter satisfaction across modes" (Aug 2014) 26 Transportation Research 160–170.

Stafford, T. "The psychology of why cyclists enrage car drivers" (12 Feb 2013) BBC News.

Statistics Canada, "Journey to work: Key results from the 2016 Census" (29 Nov 2017)—Statistics that found median distance of 7.7 km.

Teschke, K, et al. "Bicycling: Health risk or benefit?" (Mar 2012) 3(2) UBC Medical Journal.

Tomer, A. "America's commuting choices: 5 major takeaways from 2016 census data" (3 Oct 2017) Brookings.

Turcotte, M. "Like commuting? Workers' perceptions of their daily commute" (2005) Canadian Social Trends, Statistic Canada Catalogue No. 11-008—Cyclists are the only commuters who love their commute.

University of British Columbia, "Safety and travel mode" (26 Oct 2017)—This resource provides stats comparing injuries between driving, walking, and cycling.

Van Bekkum, JE, et al. "Cycle commuting and perceptions of barriers: Stages of change, gender and occupation" (18 Oct 2011) 111(6) Health Education 476–497.

Winters, M, et al. "Motivators and deterrents of bicycling: Comparing influences on decisions to ride" (2011) 38(1) Transportation 15—Good review of some of the influences, noting that safety is near the top.

Wright, R, et al. "Is urban cycling worth the risk?" (2 Sep 2016) Financial Times Magazine—Stats on deaths and injuries associated with cycling.

Zijlema, WL, et al. "Active commuting through natural environments is associated with better mental health: Results from the PHENOTYPE project" (Dec 2018) 121 Environment International 721–727.

Zwald, ML, et al. "Trends in active transportation and associations with cardiovascular disease risk factors among U.S. adults, 2007–2016" (Dec 2018) 116 Preventive Medicine 150–156—"Active transportation was inversely associated with CVD risk in recent years."

Parking

Cassady, CR, et al. "A probabilistic approach to evaluate strategies for selecting a parking space" (Feb 1998) 32 Transportation Science 30—Mathematical modeling that shows picking first spot almost always best.

Cookson, G, et al. "The impact of parking pain in the US, UK and Germany" (Jul 2017)

INRIX Research—Study that outlines the cost of wasted time parking and the hours spent searching for a spot (107 in NYC). Also, stat that says one-third in confrontation over parking.

Morgan, J. "Half of Britain's drivers suffer stress due to parking, study finds" (27 Sep 2017) Evening Standard.

Pawlowski, A. "Avoid parking rage on Black Friday! How to find a spot every time" (16 Sep 2016) Today—Commentary on strategy, with quote from Andrew Velkey.

Ruback, RB, et al. "Territorial defense in parking lots: Retaliation against waiting drivers" (1997) 27 Journal of Applied Social Psychology 821–834.

Start work

Pope, NG. "How the time of day affects productivity: Evidence from school schedules" (2016) 98(1) Review of Economics and Statistics 1–11—"A morning math class increases state test scores by an amount equivalent to increasing teacher quality by one-fourth standard deviation . . . Rearranging school schedules can lead to increased academic performance."

Public toilet seat

Doyle, P. "Footballers warned that spitting could help spread swine flu" (27 Oct 2009) Guardian—"'Spitting is disgusting at all times,' a spokesperson from the Health Protection Agency said, "'It's unhygienic and unhealthy, particularly if you spit close to other people. Footballers, like the rest of us, wouldn't spit indoors so they shouldn't do it on the football pitch.'"

Johnson, DL, et al. "Lifting the lid on toilet plume aerosol: A literature review with suggestions for future research" (Mar 2013) 41 American Journal of Infection Control 254–258.

Kafer, A. "Other people's shit (and pee!)" (2016) 115(4) South Atlantic Quarterly—Analysis of the issue of dirty toilet seats in gender-neutral toilets.

Lai, ACK, et al. "Emission strength of airborne pathogens during toilet flushing" (2018) 28(1) Indoor Air 73–79—Study exploring toilet plume.

Moore, KH, et al. "Crouching over the toilet seat: Prevalence among British gynaecological outpatients and its effect upon micturition" (Jun 1991) 98(6) British Journal of Obstetrics and Gynaecology 569–572.

Palmer, MH, et al. "Self-reported toileting behaviors in employed women: Are they associated with lower urinary tract symptoms?" (Feb 2018) 37(2) Neurology and Urodynamics 37(2)—"Most habitual toileting behaviors were not associated with urinary urgency except waiting too long to urinate when at work."

Sjögren, J, et al. "Toileting behavior and urinary tract symptoms among younger women" (Nov 2017) 28(11) International Urogynecology Journal 1677–1684.

World Health Organization, "2.1 billion people lack safe drinking water at home, more than twice as many lack safe sanitation" (21 Jul 2017).

Wash hands

Allen, L. "Science confirms the obvious: People wash their hands more when they're watched" (15 Oct 2009).

Azor-Martínez, E, et al. "Effectiveness of a multifactorial handwashing program to reduce school absenteeism due to acute gastroenteritis" (Feb 2014) 33(2) Pediatric Infectious Disease Journal 34–39—"The school children from the EG [experimental group] had a 36% lower risk of absenteeism due to AGE [acute gastroenteritis]."

Bearman, G, et al. "Healthcare personnel attire in non-operating-room settings" (Feb 2014) 35(2) Infection Control & Hospital Epidemiology—"The role of attire in cross-transmission remains poorly established, and until more definitive information exists priority should be placed on evidence-based measures to prevent healthcare-associated infections."

Borchgrevink, CP, et al. "Hand washing practices in a college town environment" (Apr 2013) 75(8) Journal of Environmental Health—"Only 5% or so spent more than 15 seconds in combined washing, rubbing, and rinsing of their hands."

Bradley Corporation, "Global handwashing day focuses on need for universal hand hygiene" (14 Oct 2015) PR Newswire—"While 92 percent of Americans believe it's important to wash their hands after using a public restroom, only 66 percent of Americans say they always wash up after using a public restroom. As for washing with soap, nearly 70 percent admit they've skipped the suds and just rinsed with water."

Burton, M, et al. "The effect of handwashing with water or soap on bacterial contamination of hands" (Jan 2011) 8(1) International Journal of Environmental Research and Public Health 97–104.

Centers for Disease Control and Prevention, "Hygiene fast facts" (26 Jul 2016)—"It is estimated that washing hands with soap and water could reduce diarrheal disease-associated deaths by up to 50%."

Centers for Disease Control and Prevention, "Why wash your hands?" (17 Sep 2018)—Nice analysis of the benefits of handwashing.

Curtis, V, et al. "Effect of washing hands with soap on diarrhoea risk in the community: A systematic review" (May 2003) 3(5) The Lancet: Infectious Diseases 275–281—"We extrapolate the potential number of diarrhoea deaths that could be averted by handwashing at about a million (1.1 million, lower estimate 0.5 million, upper estimate 1.4 million)."

Edmond, M. "Bare below the elbow and implications for infection control" (2017) Infectious Disease Advisor—Review of data on value of "bare below the elbow" strategies. Finds no evidence, only biological plausibility.

Huang, C, et al. "The hygienic efficacy of different hand-drying methods: A review of the evidence" (Aug 2012) 87(8) Mayo Clinic Proceedings 791–798.

Huesca-Espitia, LDC, et al. "Deposition of bacteria and bacterial spores by bathroom hot-air hand dryers" (9 Feb 2018) Applied and Environmental Microbiology.

Pickering, AJ, et al. "Video surveillance captures student hand hygiene behavior, reactivity to observation, and peer influence in Kenyan primary schools" (27 Mar 2014) 9 PLoS ONE—"Video surveillance documented higher hand cleaning rates (71%) when at least one other person was present at the hand cleaning station, compared to when a student was alone."

Spector, D. "Here's what happens if you never wash your jeans" (27 Jul 2016) Independent.

Zielinski, S. "The myth of the frozen jeans" (7 Nov 2011) Smithsonian.

Multitask

Carrier, LM, et al. "Productivity in peril: Higher and higher rates of technology multitasking" (8 Jan 2018) Behavioral Scientist.

"The digital native is a myth" (27 Jul 2017) Nature.

Hills, TT. "The dark side of information proliferation" (29 Nov 2018) 14(3) Perspectives on Psychological Science—Useful analysis (provided to me by author) of impact of cognitive selection biases.

Huth, S. "Employees waste 759 hours each year due to workplace distractions" (22 Jun 2015) Telegraph.

Kirschner, PA, et al. "The myths of the digital native and the multitasker" (Oct 2017) 67 Teaching and Teacher Education 135–142—"Information-savvy digital natives do not exist."

Sanbonmatsu, DM, et al. "Who multi-tasks and why? Multi-tasking ability, perceived multi-tasking ability, impulsivity, and sensation seeking" (2013) 8(1) PLoS ONE—"Multi-tasking activity as measured by the Media Multitasking Inventory and self-reported cell phone usage while driving were negatively correlated with *actual* multi-tasking ability. Multi-tasking was positively correlated with participants' *perceived* ability to multi-task ability which was found to be significantly inflated."

Schulte, B. "Work interruptions can cost you 6 hours a day. An efficiency expert explains how to avoid them" (1 Jun 2015) Washington Post—Article notes research on interruptions occurring every 3 minutes and 5 seconds and that it takes 23 minutes to reengage.

Shamsi, T, et al. "Disruption and recovery of computing tasks: Field study, analysis, and directions" (2007) Proceedings of the 2007 CHI Conference on Human Factors in Computing Systems 677–686.

Smith, N. "Distracted Workers Are Costing You Money" (2010) Business News Daily—"Distracted workers cost U.S. businesses $650 billion a year."

Stoet, G, et al. "Are women better than men at multi-tasking?" (2013) 1(18) BMC Psychology.

Sullivan, B, et al. "Brain, Interrupted" (3 May 2013) New York Times—Report on study of 136 participants being subjected to multitasking.

Watson, JM, et al. "Supertaskers: Profiles in extraordinary multitasking ability" (Aug 2010) 17(4) Psychonomic Bulletin & Review 479–485—"Whereas the vast majority of participants showed significant performance decrements in dual-task conditions (compared with single-task conditions for either driving or OSPAN tasks), 2.5% of the sample showed absolutely no performance decrements."

Westbrook, JI, et al. "Task errors by emergency physicians are associated with interruptions, multitasking, fatigue and working memory capacity: A prospective, direct observation study" (Jan 2018) 27(8) BMJ Quality & Safety—"Interruptions, multitasking and poor sleep were associated with significantly increased rates of prescribing errors among emergency physicians."

Anticipate lunch

Malkoc, S. "Want to be more productive? Stop scheduling out your day" (25 Jun 2018) Quartz.

Tonietto, GN, et al. "When an hour feels shorter: Future boundary tasks alter consumption by contracting time" (19 May 2018) 45(5) Journal of Consumer Research.

PART II: AFTERNOON

Lunch

American Institute for Cancer Research, "New Report Finds Whole Grains Lower Colorectal Cancer Risk" (Sep 2017).

Amirikian, K, et al. "Effects of the gluten-free diet on body mass indexes in pediatric celiac patients" (Mar 2019) 68(3) Journal of Pediatric Gastroenterology and Nutrition.

Boseley, S. "Extreme dieters eating gluten-free food alongside smoking and vomiting" (18 Jun 2018) Guardian—"Some people are unnecessarily avoiding gluten because they think it will help them lose weight."

Choung, RS, et al. "Less hidden celiac disease but increased gluten avoidance without a diagnosis in the United States: Findings from the National Health and Nutrition Examination Surveys from 2009 to 2014" (2017) 92(1) Mayo Clinic Proceedings 30–38.

Christoph, MJ, et al. "Who values gluten-free? Dietary intake, behaviors, and sociodemographic characteristics of young adults who value gluten-free food" (Aug 2018) 118(8) Journal of Academy of Nutrition and Dietetics 1389–1398.

Elliott, C. "The nutritional quality of gluten-free products for children" (Aug 2018) 142(2) Pediatrics—"Not nutritionally superior to regular child-targeted foods and may be of greater potential concern because of their sugar content."

Fernan, C, et al. "Health halo effects from product titles and nutrient content claims in the context of 'protein' bars" (30 Aug 2017) 33(12) Health Communication 1–9.

Food Insight, "Survey: Nutrition Information Abounds, But Many Doubt Food Choices" (5 May 2019) International Food Information Council Foundation—"More than half of those (56 percent) say the conflicting information makes them doubt the choices they make."

"Gluten-free: Simply a way to healthier eating?" (7 Nov 2017) Hartman Group Newsletter.

"'Gluten free' claims in the marketplace" (10 Nov 2017) Agriculture and Agri-Food Canada—22 percent of Canadians are gluten-avoiders.

Halmos, EP, et al. "Gluten in 'gluten-free' manufactured foods in Australia: A cross-sectional study" (Aug 2018) 209(10) Medical Journal of Australia.

King, J. "Experiences of coeliac disease in a changing gluten-free landscape" (2 Oct 2018) 32(1) Journal of Human Nutrition and Dietetics.

Lebwohl, B, et al. "Coeliac Disease" (28 Jul 2017) 391(10115) The Lancet 70–81.

Lebwohl, B, et al. "Long term gluten consumption in adults without celiac disease and risk of coronary heart disease: Prospective cohort study" (2 May 2017) BMJ 357—"The promotion of gluten-free diets among people without celiac disease should not be encouraged."

Lis, D, et al. "Exploring the popularity, experiences, and beliefs surrounding gluten-free diets in nonceliac athletes" (Feb 2015) 25(1) International Journal of Sport Nutrition and Exercise Metabolism 37–45.

Lis, D, et al. "No effects of a short-term gluten-free diet on performance in nonceliac athletes" (Dec 2015) 47(12) Medicine & Science in Sports & Exercise 2563–2570.

McFadden, B. "'Gluten-free water' shows absurdity of trend in labeling what's absent" (28 Aug 2017) The Conversation.

Meeting News, "Most who avoid gluten lack symptoms of intolerance, sensitivity" (5 Jun 2018) Healio.

Niland, B, et al. "Health benefits and adverse effects of a gluten-free diet in non–celiac disease patients" (Feb 2018) 14(2) Gastroenterology & Hepatology—"Gluten avoidance may be associated with adverse effects in patients without proven gluten-related diseases."

Okada, EM. "Justification effects on consumer choice of hedonic and utilitarian goods" (Feb 2005) 42(1) Journal of Marketing Research 43–53.

Prada, M, et al. "The impact of a gluten-free claim on the perceived healthfulness, calories, level of processing and expected taste of food products" (Apr 2019) 73 Food Quality and Preference—"GF products were perceived as more healthful."

Saplakoglu, Y. "Keto, Mediterranean or vegan: Which diet is best for the heart?" (12 Nov 2018) Live Science.

Servick, K. "What's really behind 'gluten sensitivity'?" (23 May 2018) Science.

Stevens, L. "Gluten-free and regular foods: A cost comparison" (Aug 2008) 69(3) Canadian Journal of Dietetic Practice and Research 147–150—"On average, gluten-free products were 242% more expensive than regular products."

Rant!

Bogart, N. "1 in 4 young people regret posts on social media" (29 Jul 2013) Global News.

Bushman, BJ. "Does venting anger feed or extinguish the flame? Catharsis, rumination, distraction, anger, and aggressive responding" (2002) 28(6) Personality and Social Psychology Bulletin, 724–731—"Doing nothing at all was more effective than venting anger. These results directly contradict catharsis theory."

Chapman, BP, et al. "Emotion suppression and mortality risk over a 12-year follow-up" (6 Aug 2013) 75(4) Journal of Psychosomatic Research 381–385—"Emotion suppression may convey risk for earlier death, including death from cancer." Small correlation study.

Courter, B. "Studies show confession numbers are falling, but it's still considered important" (30 Jun 2014) Times Free Press—"Going to confession is often a cathartic experience."

Davidson, KW, et al. "Anger expression and risk of coronary heart disease: Evidence from the Nova Scotia Health Survey" (Feb 2010) 159(2) American Heart Journal 199–206.

Fan, R, et al. "Anger is more influential than joy: Sentiment correlation in Weibo" (15 Oct 2014) 9 PLoS ONE—"We find that the correlation of anger among users is significantly higher than that of joy."

Frey, S, et al. "The rippling dynamics of valenced messages in naturalistic youth chat" (Aug 2019) 51(4) Behavior Research Methods.

Heikkilä, K, et al. "Work stress and risk of cancer: Meta-analysis of 5700 incident cancer events in 116 000 European men and women" (7 Feb 2013) 346 BMJ—"Work related stress, measured and defined as job strain, at baseline is unlikely to be an important risk factor for colorectal, lung, breast, or prostate cancers."

Kye, SY, et al. "Perceptions of cancer risk and cause of cancer risk in Korean adults" (Apr 2015) 47(2) Cancer Research and Treatment 158–165—"The most important perceived cause of cancer risk was stress."

Leung, FH, et al. "Bless me, for I have sinned . . . Behaviour change and the confessional" (Jan 2009) 55(1) Canadian Family Physician 17–18—"Even without delving into the theology of confession, its cathartic nature is evident."

Martin, RC. "Three facts about venting online" (1 Aug 2014) Psychology Today—"46 percent of Twitter users say they often tweet as a way of dealing with or venting anger."

Martin, RC, et al. "Anger on the Internet: The perceived value of rant-sites" (Feb 2013) 16(2) Cyberpsychology, Behavior, and Social Networking—For most participants, reading and writing rants were associated with negative shifts in mood.

Mostofsky, E, et al. "Outbursts of anger as a trigger of acute cardiovascular events: A systematic review and meta-analysis" (1 Jun 2014) 35(21) European Heart Journal 1404–1410—"There is a higher risk of cardiovascular events shortly after outbursts of anger."

Mostofsky, E, et al. "Relation of outbursts of anger and risk of acute myocardial infarction" (1 Aug 2013) 112(3) American Journal of Cardiology 343–348—"The risk of experiencing AMI [acute myocardial infarction] was more than twofold greater after outbursts of anger compared with at other times, and greater intensities of anger were associated with greater relative risks."

Neil, SP. "More than half of Americans have social media regrets" (6 Dec 2017) HuffPostl—Summary of study by YouGov Omnibus that found 57 percent regret a post.

"Psychological stress and cancer" (10 Dec 2012) National Cancer Institute.

Shahab, L, et al. "Prevalence of beliefs about actual and mythical causes of cancer and their association with socio-demographic and health-related characteristics: Findings from a cross-sectional survey in England" (26 Apr 2018) 103 European Journal of Cancer—43 percent believe, wrongly, that stress causes cancer.

"Stress" (11 May 2018) Cancer Research UK—"Most scientific studies have found that stress does not increase the risk of cancer."

Thank-you letter

Algoe, SB. "Putting the 'you' in 'thank you': Examining other-praising behavior as the active relational ingredient in expressed gratitude" (7 Jun 2016) 7(7) Social Psychology and Personality Science, 658–666.

Kumar, A, et al. "Undervaluing gratitude: Expressers misunderstand the consequences of showing appreciation" (27 Jun 2018) 29(9) Psychological Science 1–13.

Stand up

Adams, J, et al. "Why are some population interventions for diet and obesity more equitable and effective than others? The role of individual agency" (5 Apr 2016) 13(4) PLoS Medicine—"Population interventions that require individuals to use a high level of agency to benefit tend to be favoured by governments around the world."

Atkins, JD. "Inactivity induces resistance to the metabolic benefits following acute exercise" (1 Apr 2019) 126(4) Journal of Applied Physiology—"These data indicate that physical inactivity (e.g., sitting ~13.5 h/day and <4,000 steps/day) creates a condition whereby people become 'resistant' to the metabolic improvements that are typically derived from an acute bout of aerobic exercise (i.e., exercise resistance)."

Baker, R, et al. "A detailed description of the short-term musculoskeletal and cognitive effects of prolonged standing for office computer work" (7 Feb 2018) 61(7) Ergonomics 877–890—"The observed changes suggest replacing office work sitting with standing should be done with caution . . . In a laboratory study involving 2 h prolonged standing discomfort increased (all body areas), reaction time and mental state deteriorated while creative problem-solving improved. Prolonged standing should be undertaken with caution."

Betts, J, et al. "The energy cost of sitting versus standing naturally in man" (Apr 2019) 51(4) Medicine & Science in Sports & Exercise—"Interventions designed to reduce sitting typically encourage 30 to 120 min more standing in situ (rather than perambulation [movement]), so the 12% difference from sitting to standing reported here does not represent an effective strategy for the treatment of obesity (i.e., weight loss)."

Buckley, JP, et al. "The sedentary office: An expert statement on the growing case for change towards better health and productivity" (2015) 49(21) British Journal of Sports Medicine.

Chau, JY, et al. "Overselling sit-stand desks: News coverage of workplace sitting guidelines" (Dec 2018) 33(12) Health Communication 1475–1481.

Cheval, B, et al. "Avoiding sedentary behaviors requires more cortical resources than avoiding physical activity: An EEG study" (Oct 2018) 119 Neuropsychologia 68–80—"Additional brain resources are required to escape a general attraction toward sedentary behaviors."

Duvivier, B, et al. "Reducing sitting time versus adding exercise: Differential effects on biomarkers of endothelial dysfunction and metabolic risk" (5 Jun 2018) 8(1) Scientific Reports—Study found sitting not great for metabolism. "Light physical activity and moderate-to-vigorous physical activity had a differential effect on risk markers of cardio-metabolic health and suggest the need of both performing structured exercise as well as reducing sitting time on a daily basis."

Edwardson, CL, et al. "Effectiveness of the Stand More AT (SMArT) Work intervention: Cluster randomised controlled trial" (8 Aug 2018) 363 BMJ.

Gray, C. "Reducing sedentary behaviour in the workplace" (2018) 363 BMJ—"Questions remain about the SMArT Work intervention's transferability beyond the National Health Service."

Hanna, F, et al. "The relationship between sedentary behavior, back pain, and psychosocial correlates among university employees" (9 Apr 2019) 7 Frontiers in Public Health—"These findings suggest that sedentary employees are exposed to increasing occupational hazards such as back pain and mental health issues."

Júdice, PB, et al. "What is the metabolic and energy cost of sitting, standing and sit/stand transitions?" (Feb 2016) 116(2) European Journal of Applied Physiology 263–273.

"Key Statistics for Colorectal Cancer" (24 Jan 2019) American Cancer Society—"Overall, the lifetime risk of developing colorectal cancer is: about 1 in 22 (4.49%) for men and 1 in 24 (4.15%) for women."

MacEwen, BT, et al. "Sit-stand desks to reduce workplace sitting time in office workers with abdominal obesity: A randomized controlled trial" (Sep 2017) 14(9) Journal of Physical Activity and Health 710–715—"Sit-stand desks were effective in reducing workplace sedentary behavior in an at-risk population, with no change in sedentary behavior or physical activity outside of work hours. However, these changes were not sufficient to improve markers of cardiometabolic risk in this population."

Mansoubi, M, et al. "Using sit-to-stand workstations in offices: Is there a compensation effect?" (Apr 2016) 48(4) Medicine & Science in Sports & Exercise 720–725—"These changes were compensated for by reducing activity and increasing sitting outside of working hours."

Mantzari, E, et al. "Impact of sit-stand desks at work on energy expenditure, sitting time and cardio-metabolic risk factors: Multiphase feasibility study with randomised controlled component" (Mar 2019) 13 Preventive Medicine Reports—"The overall effect of sit-stand desks for reducing sitting at work is uncertain . . . Preliminary evidence suggests the desks' potential to reduce workplace sitting but raises concern about their potential to adversely affect energy expenditure and sitting time outside work."

Patel, AV, et al. "Prolonged leisure time spent sitting in relation to cause-specific mortality in a large US cohort" (1 Oct 2018) 187(10) American Journal of Epidemiology—Study found that prolonged sitting was associated with higher risk of mortality from a range of causes, including cancer and heart disease.

"The Price of Inactivity" (2015) American Heart Association.

Riotta, C. "Standing at work is just as unhealthy as smoking cigarettes daily, study says" (9 Sep 2017) Newsweek.

Shrestha, N, et al. "Workplace interventions (methods) for reducing time spent sitting at work" (20 Jun 2018) Cochrane Library—"At present there is low-quality evidence that the use of sit-stand desks reduce workplace sitting at short-term and medium-term follow-ups. However, there is no evidence on their effects on sitting over longer follow-up periods."

Smith, P, et al. "The relationship between occupational standing and sitting and incident heart disease over a 12-year period in Ontario, Canada" (1 Jan 2018) 187(1) American Journal of Epidemiology 27–33—Study of over 7,000 employees found that "occupations involving predominantly standing were associated with an approximately 2-fold risk of heart disease compared with occupations involving predominantly sitting."

Snowbeck, C. "Standing desks have become an important workplace benefit" (15 Sep 2017) Waterloo Region Record—"Standing desks have emerged as the fastest growing employee benefit in U.S. workplaces, according to a June report from the Society for Human Resource Management."

Stamatakis, E, et al. "Sitting behaviour is not associated with incident diabetes over 13 years: The Whitehall II cohort study" (May 2017) 51(10) British Journal of Sports Medicine 818–823—"We found limited evidence linking sitting and incident diabetes over 13 years in this occupational cohort of civil servants."

Vallance, JK, et al. "Evaluating the evidence on sitting, smoking, and health: Is sitting really the new smoking?" (Nov 2018) 108(11) American Journal of Public Health 1478–1482.

Ward, R. "Sitting may be bad but it's still better than smoking, Alberta researcher says" (1 Oct 2018) CBC News.

Wilcken, H. "Is sitting the new smoking, or isn't it?" (25 Sep 2017) 37 Medical Journal of Australia.

Yang, L. "Trends in sedentary behavior among the US population, 2001–2016" (30 Apr 2019) 321(16) JAMA—"The estimated prevalence of sitting watching television or videos at least 2 h/d was high in 2015-2016 (ranging from 59% to 65%); the estimated prevalence of computer use outside school or work for at least 1 h/d increased from 2001 to 2016."

Another coffee?

Haber, N, et al. "Causal language and strength of inference in academic and media articles shared in social media (CLAIMS): A systematic review" (30 May 2018) 13(5) PLoS ONE—"We find a large disparity between the strength of language as presented to the research consumer and the underlying strength of causal inference among the studies most widely shared on social media."

Loftfield, E, et al. "Association of coffee drinking with mortality by genetic variation in caffeine metabolism findings from the UK biobank" (2 Jul 2018) 178(8) JAMA—"This study provides further evidence that coffee drinking [including among those who drink 8 or more cups a day] can be part of a healthy diet and offers reassurance to coffee drinkers."

Nagler, RH. "Adverse outcomes associated with media exposure to contradictory nutrition messages" (2014) 19(1) Journal of Health Communication 24–40—"Exposure to conflicting information on the health benefits and risks of, for example, wine, fish, and coffee consumption is associated with confusion about what foods are best to eat and the belief that nutrition scientists keep changing their minds. There is evidence that these beliefs, in turn, may lead people to doubt nutrition and health recommendations more generally."

Rettner, R. "Here's how much caffeine you need, and when, for peak alertness" (6 Jun 2018) Live Science.

Selvaraj, S, et al. "Media coverage of medical journals: Do the best articles make the news?" (17 Jan 2014) 9(1) PLoS ONE—"Newspapers were more likely to cover observational studies and less likely to cover RCTs [randomized controlled trials] than high impact journals. Additionally, when the media does cover observational studies, they select articles of inferior quality. Newspapers preferentially cover medical research with weaker methodology."

Victory, J. "A venti-sized serving of misinformation in news stories on latest coffee study" (3 Jul 2018) HealthNewsReview.

Vital-Lopez, FG, et al. "Caffeine dosing strategies to optimize alertness during sleep loss" (28 May 2018) 27(5) Journal of Sleep Research—Research by the U.S. Department of Defense on when to drink coffee.

Wang, M, et al. "Reporting of limitations of observational research" (8 Jun 2015) 175(9) JAMA International Medicine 1571–1572—"Limitations of observational research published in high-impact journals were infrequently mentioned in associated news stories."

Soap

Bannan, EA, et al. "The inability of soap bars to transmit bacteria" (Jun 1965) 55(6) American Journal of Public Health and the Nation's Health 915–922.

Burton, M, et al. "The effect of handwashing with water or soap on bacterial contamination of hands" (Jan 2011) 8(1) International Journal of Environmental Research and Public Health 97–104—"Handwashing with non-antibacterial soap and water is more effective for the removal of bacteria of potential faecal origin from hands than handwashing with water alone."

Centers for Disease Control and Prevention, "When & how to use hand sanitizer" (17 Sep 2019).

Centers for Disease Control and Prevention, "When & how to wash your hands" (Sep 18 2019).

Food and Drug Administration, "Antibacterial soap? You can skip it, use plain soap and water" (16 May 2019)—"If you use these products because you think they protect you more than soap and water, that's not correct."

Food and Drug Administration, "FDA issues final rule on safety and effectiveness of antibacterial soaps" (2 Sep 2016).

Heinze, JE, et al. "Washing with contaminated bar soap is unlikely to transfer bacteria" (Aug 1988) 101(1) Epidemiology & Infection 135–142—"These findings, along with other published reports, show that little hazard exists in routine handwashing with previously used soap bars and support the frequent use of soap and water for handwashing to prevent the spread of disease." Although this study was funded by the soap industry, the data has not been contradicted.

Luby, SP, et al. "Effect of intensive handwashing promotion on childhood diarrhea in high-risk communities in Pakistan: A randomized controlled trial" (2 Jun 2004) 291(21) JAMA 2547–2554—"Improvement in handwashing in the household reduced the incidence of diarrhea among children at high risk of death from diarrhea."

Pickering, AJ, et al. "Efficacy of waterless hand hygiene compared with handwashing with soap: A field study in Dar es Salaam, Tanzania" (Feb 2010) 82(2) Tanzania American Journal of Tropical Medicine and Hygiene 270–278—"Hand sanitizer was significantly better than handwashing with respect to reduction in levels of fecal streptococci."

Ruffins, E. "Recycling hotel soap to save lives" (16 Jun 2011) CNN—Story on Derreck Kayongo.

Zapka, CA, et al. "Bacterial hand contamination and transfer after use of contaminated bulk-soap-refillable dispensers" (May 2011) 77(9) Applied and Environmental Microbiology 2898–2904—"Washing with contaminated soap from bulk-soap-refillable dispensers can increase the number of opportunistic pathogens on the hands."

Drink water

"Bottled water contains more bacteria than tap water" (25 May 2010) Telegraph.

Clark, WF, et al. "Effect of coaching to increase water intake on kidney function decline in adults with chronic kidney disease" (8 May 2018) 319(18) JAMA 1870–1879.

Fenton, T, et al. "Systematic review of the association between dietary acid load, alkaline water and cancer" (Jun 2016) 6(6) BMJ Open.

Piantadosi, C. "'Oxygenated' water and athletic performance" (Sep 2006) 40(9) British Journal of Sports Medicine 740–741.

Rosinger, A, et al. "Association of caloric intake from sugar-sweetened beverages with water intake among US children and young adults in the 2011–2016 National Health and Nutrition Examination Survey" (1 Jun 2019) 173(6) JAMA Pediatrics.

Schwarcz, J. "Alkaline water nonsense" (20 Mar 2017) McGill Office for Science and Society.

"Taste test: Is bottled water better than tap?" (22 Mar 2012) CTV Atlantic.

Williams-Grut, O. "People prefer tap water over 'premium' £1.49 Fiji Water in a blind taste test" (14 May 2017) Business Insider.

Office meeting

Allen, JA, et al. "Let's get this meeting started: Meeting lateness and actual meeting outcomes" (24 Mar 2018) 39(8) Journal of Organizational Behavior 1008–1021.

Association for Psychological Sciences, "There's a Better Way to Brainstorm" (15 Mar 2016).

Bernstein, ES, et al. "The impact of the 'open' workspace on human collaboration" (2 Jul 2018) 373(1753) Philosophical Transactions of the Royal Society—70 percent decrease in face-to-face interaction.

Brown, VR, et al. "Making group brainstorming more effective: Recommendations from an associative memory perspective" (1 Dec 2002) 11(6) Current Directions in Psychological Science—"Much literature on group brainstorming has found it to be less effective than individual brainstorming."

Carey, B. "Can big science be too big?" (13 Feb 2019) New York Times.

Chamorro-Premuzic, T. "Why group brainstorming is a waste of time" (25 Mar 2015) Harvard Business Review—"Ultimately, brainstorming continues to be used because it feels intuitively right to do so."

Derdowski, LA. "Here's why you dread brainstorming at work" (14 Nov 2018) Medical Xpress—"Social processes in working groups can effectively prevent teammates from achieving the desired state of the creative synergy. Clearly, group interactions interfere with the very advantage they are expected to provide."

Furnham, A. "The brainstorming myth" (6 Jan 2003) 11(4) Business Strategy Review 21–28—"Research shows unequivocally that brainstorming groups produce fewer and poorer quality ideas than the same number of individuals working alone."

Greenwood, V. "Is conference room air making you dumber?" (6 May 2019) New York Times.

Kauffield, S, et al. "Meetings matter: Effects of team meetings on team and organizational success" (Apr 2012) 43(2) Small Group Research 130–158.

Khazanchi, S, et al. "A spatial model of work relationships: The relationship-building and relationship-straining effects of workspace design" (11 Oct 2018) 43(4) Academy of Management Review.

Lehrer, J. "Brainstorming: An idea past its prime" (19 Apr 2012) Washington Post—"Decades of research have consistently shown that brainstorming groups think of far fewer ideas than the same number of people who work alone and later pool their ideas."

Lehrer, J. "Groupthink" (30 Jan 2012) New Yorker—Explores the brainstorming myth.

Microsoft, "Survey finds workers average only three productive days per week" (15 Mar 2005)—Worldwide, "people spend 5.6 hours each week in meetings; 69 percent feel meetings aren't productive (U.S.: 5.5 hours; 71 percent feel meetings aren't productive)."

Perlow, LA, et al. "Stop the Meeting Madness" (Jul–Aug 2017) Harvard Business Review.

Rogelberg, SG, et al. "The science and fiction of meetings" (Dec 2007) 48(2) MIT Sloan Management Review—"Conservatively, the average employee spends approximately six hours per week in scheduled meetings."

Romano, NC, et al. "Meeting analysis: Findings from research and practice" (Feb 2001) Proceedings of the 34th Hawaii International Conference on System Sciences—"Self estimates of meeting productivity by managers in many different functional areas range from 33% - 47%."

Thompson, D. "Study: Nobody is paying attention on your conference call" (14 Aug 2014) Atlantic—Summarizes data on how no one is paying attention.

Tonietto, GN, et al. "When an hour feels shorter: Future boundary tasks alter consumption by contracting time" (19 May 2018) 45(5) Journal of Consumer Research—Study demonstrated that a looming meeting makes us less productive and feel that time goes by faster.

Ward, T. "Walking meetings? Proceed with caution" (13 Mar 2017) Psychology Today—Nice review of the science, which shows mixed results.

Yoerger, M, et al. "The impact of premeeting talk on group performance" (2018) 49 Small Group Research 226–258.

Nap time

Bonnar, D, et al. "Sleep interventions designed to improve athletic performance and recovery: A systematic review of current approaches" (Mar 2018) 48(3) Sports Medicine 683–703.

Brooks, A, et al. "A brief afternoon nap following nocturnal sleep restriction: Which nap duration is most recuperative?" (Jun 2006) 29(6) Sleep—"The 10-minute nap was overall the most effective afternoon nap duration of the nap lengths examined in this study."

Chen, GC, et al. "Daytime napping and risk of type 2 diabetes: A meta-analysis of prospective studies" (Sep 2018) 22(3) Sleep and Breathing—"This meta-analysis suggests that daytime napping is associated with an increased risk of T2D." Because of limited good studies, "our findings should be interpreted with great caution."

Cheungpasitporn, W, et al. "The effects of napping on the risk of hypertension: A systematic review and meta-analysis" (Nov 2016) 9(4) Journal of Evidence-Based Medicine—"Our meta-analysis demonstrates a significant association between daytime napping and hypertension."

DeMers, J. "Will you actually be more productive if you take a nap every day?" (5 Jun 2017) Entrepreneur—Example of an article that refers to a pilot study without noting it was a small one.

Fan, F, et al. "Daytime napping and cognition in older adults" (27 Apr 2018) 41(1) Sleep—"Nap within a certain duration/frequency may be protective for cognitive function. Overall, nap interventions demonstrated favorable cognitive effects."

Goldman, SE, et al. "Association between nighttime sleep and napping in older adults" (May

2008) 31(5) Sleep 733–740—"More sleep fragmentation was associated with higher odds of napping although not with nap duration."

Goldschmied, JG, et al. "Napping to modulate frustration and impulsivity: A pilot study" (12 Jun 2015) 86 Personality and Individual Differences 164–167—Pilot study used to support the idea that napping reduces frustration, etc.

Guo, VY, et al. "The association between daytime napping and risk of diabetes: A systematic review and meta-analysis of observational studies." (18 Jan 2017) 37 Sleep Medicine 105–112—"Long daytime napping over 1 h per day was associated with increased risk of both prevalent and incident DM [diabetes mellitus]. Further studies are needed to confirm the findings."

Heffron, TM. "Insomnia Awareness Day facts and stats" (10 Mar 2014).

Hilditch, CJ, et al. "A 30-minute, but not a 10-minute nighttime nap is associated with sleep inertia" (1 Mar 2016) 39(3) Sleep 675–685.

Hublin, C, et al. "Napping and the risk of type 2 diabetes: A population-based prospective study" (Jan 2016) 17 Sleep Medicine 144–148.

Makino, S, et al. "Association between nighttime sleep duration, midday naps, and glycemic levels in Japanese patients with type 2 diabetes" (Apr 2018) 44 Sleep Medicine 4–11—"Poor sleep quality and quantity could aggravate glycemic control in type 2 diabetes. Midday naps could mitigate the deleterious effects of short nighttime sleep on glycemic control."

Mantua, J, et al. "Exploring the nap paradox: Are mid-day sleep bouts a friend or foe?" (Sep 2017) 37 Sleep Medicine 88–97—"In older populations, as opposed to the obvious benefits of a mid-day nap outlined above, excessive napping has been linked with negative outcomes."

McVeigh, T. "Insomnia: Britons' health 'at risk' as 50% fail to get enough sleep" (13 Nov 2011) Guardian.

Milner, CE, et al. "Benefits of napping in healthy adults: Impact of nap length, time of day, age, and experience with napping" (Jun 2009) 18(2) Journal of Sleep Research 272–281—Review of variables that affect whether napping is beneficial. "The existing literature shows that certain variables, such as the timing and duration of a nap, age, and experience with napping are important moderators of the benefits of naps."

Mitler, MM, et al. "Catastrophes, sleep, and public policy: Consensus report" (Feb 1988) 11(1) Sleep 100–109—Notes that industrial accidents are associated with sleepiness, including Chernobyl and the *Challenger* disasters.

National Sleep Foundation, "Napping"—"If you have trouble sleeping at night, a nap will only amplify problems."

National Sleep Foundation, "The relationship between sleep and industrial accidents."

"One in four Americans develop insomnia each year: 75 percent of those with insomnia recover" (5 Jun 2018) ScienceDaily.

Owens, JF, et al. "Napping, nighttime sleep, and cardiovascular risk factors in mid-life adults" (15 Aug 2010) 6(4) Journal of Clinical Sleep Medicine 330–335—"Napping in middle-aged men and women is associated with overall less nighttime sleep in African Americans and lower sleep efficiency . . . and increased BMI and central adiposity."

Petit, E, et al. "A 20-min nap in athletes changes subsequent sleep architecture but does not alter physical performances after normal sleep or 5-h phase-advance conditions" (Feb 2014) 114(2) European Journal of Applied Physiology 305–315—"Napping showed no reliable benefit on short-term performances of athletes exercising at local time or after a simulated jet lag."

Rosekind, MR, et al. "The cost of poor sleep: Workplace productivity loss and associated costs" (Jan 2010) 52(1) Journal of Occupational and Environmental Medicine 91–98—"Sleep disturbances contribute to decreased employee productivity at a high cost to employers."

Samuels, C, et al. "Sleep as a recovery tool for athletes" (17 Nov 2014) British Journal of Sports Medicine Blog.

Sleep Health Foundation, "Insomnia" (2011) http://sleephealthfoundation.org.au/pdfs/Insomnia.pdf—"Around 1 in 3 people have at least mild insomnia."

Tietzel, AJ, et al. "The short-term benefits of brief and long naps following nocturnal sleep restriction" (1 May 2001) 24(3) Sleep—"Because the delayed benefits following the 30-minute nap may be due to sleep inertia, longer post-nap testing periods should be investigated. However, we conclude that the detrimental effects of sleep restriction were more rapidly and significantly ameliorated, at least within the hour following the nap, by a 10-minute afternoon nap."

Wannamethee, SG, et al. "Self-reported sleep duration, napping, and incident heart failure: Prospective associations in the British Regional Heart Study" (Sep 2016) 64(9) Journal of the American Geriatrics Study 1845–1850—"Self-reported daytime napping of longer than 1 hour is associated with greater risk of HF [heart failure] in older men."

Watson, AM. "Sleep and athletic performance" (Nov–Dec 2017) 16(6) Current Sports Medicine Reports 413–418—"The role of daytime naps on performance is unclear."

Weir, K. "The science of naps" (Jul–Aug 2016) 47(7) American Psychological Association Monitor on Psychology.

Yamada, T, et al. "Daytime napping and the risk of cardiovascular disease and all-cause mortality: A prospective study and dose-response meta-analysis" (1 Dec 2015) 38(12) Sleep 1945–1953—"Meta-analysis demonstrated a significant J-curve dose-response relation between nap time and cardiovascular disease."

Yamada, T, et al. "J-curve relation between daytime nap duration and type 2 diabetes or metabolic syndrome: A dose-response meta-analysis" (2 Dec 2016) 6 Scientific Reports—"Dose-response meta-analysis showed a J-curve relation between nap time and the risk of diabetes or metabolic syndrome, with no effect of napping up to about 40 minutes/day, followed by a sharp increase in risk at longer nap times. In summary, longer napping is associated with an increased risk of metabolic disease. Further studies are needed to confirm the benefit of a short nap."

Five-second rule

Aston University, "Researchers prove the five-second rule is real" (10 Mar 2014)—"87% of people surveyed said they would eat food dropped on the floor, or already have done so . . . 81% of the women who would eat food from the floor would follow the 5 second rule." This is a report on one of the few studies on point but, as noted in the paper, "this research has not yet been peer-reviewed."

Beaulieu, M. "The '5-second rule' has been officially sanctioned by a germ scientist" (15 Mar 2017) CBC Life—"In a survey of over 2,000 hungry humans, an impressive 79 percent of them confessed to eating food that had hit the floor."

Discovery Channel, "5 second rule with food on floor"—Description of the *MythBusters* show on the issue.

Midkiff, S. "The percentage of people who apply the five-second rule might upset your stomach" (13 Dec 2017) Refinery29.

Miranda, RC, et al. "Longer contact times increase cross-contamination of *Enterobacter aerogenes* from surfaces to food" (2 Sep 2016) 82(21) Applied and Environmental Microbiology—One of the few rigorously done studies.

Sidder, A. "What does science say about the five-second rule? It's complicated" (13 Sep 2016) Smithsonian—"The five-second rule is a significant oversimplification of what actually happens when bacteria transfer from a surface to food," [Donald] Schaffner said. "Bacteria can contaminate instantaneously."

University of Illinois at Urbana-Champaign, "If you drop it, should you eat it? Scientists weigh in on the 5-second rule" (2 Sep 2003)—Report on high school student Jillian Clarke's unpublished study.

Email

Ariely, D. "How many of our emails should we know about the moment someone decides to email us?" (23 Feb 2017) Dan Ariely—"205 billion. That's the number of emails we sent and received in 2015, and that number is expected to grow to 246 billion by 2019 . . . As it turns out, very few—only 12%!—of emails need to be seen within 5 minutes of being sent."

Barley, SR, et al. "E-mail as a source and symbol of stress" (Jul–Aug 2011) 22(4) Organization Science 887–906—"The more time people spent handling e-mail, the greater was their sense of being overloaded, and the more e-mail they processed, the greater their perceived ability to cope."

Beck, J. "How it became normal to ignore texts and emails" (11 Jan 2018) Atlantic.

Becker, WJ, et al. "Killing me softly: Electronic communications monitoring and employee and spouse well-being" (9 Jul 2018) 2018(1) Academy of Management Proceedings—

"Detrimental health and relationship effects of expectations were mediated by negative affect. This includes crossover effects of electronic communication expectations on partner health and marital satisfaction."

Burnett, J. "Study: The average worker's inbox contains 199 unread emails" (2 Oct 2017) Ladders—Reports that 94 percent of workers "rely on email for work management."

Burnett, J. "24% of Americans think reaching 'inbox-zero' is an impossibility" (23 Aug 2018) Ladders.

"Carleton study finds people spending a third of job time on email" (20 Apr 2017) Carleton Newsroom.

Clark, D. "Why email is so stressful, even though it's not actually that time-consuming" (9 Apr 2018) Harvard Business Review.

Collins, N. "Email 'raises stress levels'" (4 Jun 2013) Telegraph—Report on study that finds email raises blood pressure. Stress levels "peaked at points in the day when people's inboxes were fullest."

Counts, V. "De-clutter your inbox: Transform your perspective to see email as a tool" (28 Sep 2017) Proceedings of the Human Factors and Ergonomics Society Annual Meeting—Study suggests that having only a few folders is the best approach.

Dewey, C. "How many hours of your life have you wasted on work email? Try our depressing calculator" (3 Oct 2016) Washington Post—"We spend an average of 4.1 hours checking our work email each day." Article also reports that 79 percent check email on vacation.

Jackson, T, et al. "Case study: Evaluating the effect of email interruptions within the workplace" (Jan 2002) Conference on empirical assessment in software engineering 3–7—"The majority of emails, 70%, were reacted to within 6 seconds of them arriving and 85% were reacted to within 2 minutes of arriving."

Jerejian, ACM, et al. "The contribution of email volume, email management strategies and propensity to worry in predicting email stress among academics" (May 2013) 29(3) Computers in Human Behavior 991–996—Email volume predicted stress and "email management did not moderate the email volume and stress relationship."

Kelleher, D. "Survey: 81% of U.S. employees check their work mail outside work hours" (20 May 2013) TechTalk—32 percent answer within 15 minutes and 23 percent within 30 minutes.

Kim, J, et al. "Technology supported behavior restriction for mitigating self interruptions in multi-device environments" (11 Sep 2017) 1(3) Proceedings of the ACM on Interactive, Mobile, Wearable and Ubiquitous Technologies—"Stress of the experimental group [was] lower despite its behavioral restriction with multi-device blocking."

Kooti, F, et al. "Evolution of conversations in the age of email overload" (2 Apr 2015) Proceedings of the 24th International Conference on World Wide Web—Study of over two million email users "found that users increased their activity as they received more emails, but not enough to compensate for the higher load. This means that as users became more overloaded, they replied to a smaller fraction of incoming emails and with shorter replies. However, their responsiveness remained intact and may even be faster."

Kushlev, K, et al. "Checking email less frequently reduces stress" (1 Feb 2015) 43 Computers in Human Behavior 220–228.

MacKay, J. "Productivity in 2017: What we learned from analyzing 225 million hours of work time" (Jan 2018) RescueTime.

Mark, G, et al. "Email duration, batching and self-interruption: Patterns of email use on productivity and stress" (7 May 2016) Proceedings of the 2016 CHI Conference on Human Factors in Computing Systems 1717–1728—"Batching email is associated with higher rated productivity with longer email duration, but despite widespread claims, we found no evidence that batching email leads to lower stress."

Mark, GJ, et al. "A pace not dictated by electrons: An empirical study of work without email" (May 2012) Proceedings of the Special Interest Group on Computer-Human Interaction—"Without email, people multitasked less and had a longer task focus, as measured by a lower frequency of shifting between windows and a longer duration of time spent working in each computer window. Further, we directly measured stress using wearable heart rate monitors and found that stress, as measured by heart rate variability, was lower without email."

Marulanda-Carter, L, et al. "Effects of e-mail addiction and interruptions on employees" (Mar 2012) 14(1) Journal of Systems and Information Technology 82–94—"E-mail interruptions have a negative time impact upon employees and show that both interrupt handling and recovery time exist. A typical task takes one third longer than undertaking a task with no e-mail interruptions."

Neporent, L. "Most emails answered in just two minutes, study finds" (Apr 13, 2015) ABC News—Nearly 90 percent of users replied to their emails within a day, with about half responding in around 47 minutes. The most frequently occurring reply time was just two minutes.

O'Donnell, B. "Most U.S. workplaces still use 'old-school' tech like email and phone calls to communicate" (22 Feb 2017) Recode.

Park, Y, et al. "The long arm of email incivility: Transmitted stress to the partner and partner work withdrawal" (8 May 2018) 39(10) Journal of Organizational Behaviour.

Patrick, VM, et al. "How to say 'no': Conviction and identity attributions in persuasive refusal" (Dec 2012) 29(4) International Journal of Research in Marketing 390–394.

Patrick, VM, et al. "'I don't'" versus 'I can't': When empowered refusal motivates goal-directed behavior" (1 Aug 2012) 39(2) Journal of Consumer Research 371–381.

Pavlus, J. "How email became the most reviled communication experience ever" (15 Jun 2015) Fast Company.

Pielot, M, et al. "Productive, anxious, lonely: 24 hours without push notifications" (4 Sep 2017) Proceedings of the 19th International Conference on Human-Computer Interaction with Mobile Devices and Services—"The evidence indicates that notifications have locked us in a dilemma: without notifications, participants felt less distracted and more productive. But, they also felt no longer able to be as responsive as expected, which made some participants anxious. And, they felt less connected with one's social group."

Reeder, B. "The best times to send email for replies (backed by data)" Yesware—"Email open and reply rates are highest on the weekends, when inbox competition is low."

Russell, E. "Strategies for effectively managing email at work" (Sep 2017)—Review concludes that it is a myth that we should check only a few times a day. But "by turning off email alerts and allocating time to check and deal with it at regular intervals, research reports that people feel more in control and less overloaded by email."

Staley, O. "Inbox Zero is a waste of time. This is how a world-class behavioral economist tames his email" (22 Mar 2017) Quartz—Dan Ariely: "Obsessively sorting and deleting emails is 'structured procrastination.'"

Stich, JF, et al. "E-mail load, workload stress and desired e-mail load: A cybernetic approach" (2019) Information Technology & People—"Higher e-mail load is associated with higher workload stress."

Tanase, L. "Email is still your customers' preferred communication tool" (Jun 2018) Entrepreneur.

"Third of Brits are so stressed they have checked work emails in middle of night, study finds" (16 May 2018) Independent.

Troy, D. "The truth about email: What's a normal inbox?" (5 Apr 2013) Pando—"Among the inboxes we sampled, the average size is 8,024 messages."

"The ultimate list of marketing statistics for 2018" (2018) HubSpot—"86% of professionals prefer to use email when communicating for business purposes."

Handshake

Bernieri, FJ, et al. "The influence of handshakes on first impression accuracy" (Apr 2011) 6(2) Social Influence 78–87.

Bishai, D, et al. "Quantifying school officials' exposure to bacterial pathogens at graduation ceremonies using repeated observational measures" (2011) 27(3) Journal of School Nursing 219–224—"We measured a risk of one new bacterial acquisition in a sample exposed to 5,209 handshakes yielding an overall estimate of 0.019 pathogens acquired per handshake. We conclude that a single handshake at a graduation offers only a small risk of bacterial pathogen acquisition."

Boshell, P. "How many hands will you shake in your lifetime?" (9 Jun 2015) Deb.

Dahl, E. "Cruise tap versus handshake: Using common sense to reduce hand contamination and germ transmission on cruise ships" (2016) 67(4) International Maritime Health 181–184.

Dolcos, S, et al. "The power of a handshake: Neural correlates of evaluative judgments in observed social interactions" (Dec 2012) 24(12) Journal of Cognitive Neuroscience 2292–2305.

Firth, J, et al. "Grip strength is associated with cognitive performance in schizophrenia and the general population: A UK Biobank study of 476559 participants" (6 Jun 2018) 44(4) Schizophrenia Bulletin 728–736.

Frumin, I, et al. "A social chemosignaling function for human handshaking" (3 Mar 2015) 4 eLife.

Ghareeb, PA, et al. "Reducing pathogen transmission in a hospital setting. Handshake versus fist bump: A pilot study" (Dec 2013) 85(4) Journal of Hospital Infection 321–323—"Implementing the fist bump in the healthcare setting may further reduce bacterial transmission between healthcare providers."

"Handshake makes for better deals in business" (3 Aug 2018) Berkeley News—Juliana Schroeder quote: "It changes the way you perceive not just the other person, but the way you frame the whole game."

LeWine, H. "Fist bump better than handshake for cleanliness" (Jul 2014) Harvard Health.

Mela, S, et al. "The fist bump: A more hygienic alternative to the handshake" (28 Jul 2014) 42(8) American Journal of Infection Control 916–917—"Adoption of the fist bump as a greeting could substantially reduce the transmission of infectious disease between individuals."

Parga, JJ, et al. "Handshake-free zone in a neonatal intensive care unit: Initial feasibility study" (1 Jul 2017) 45(7) American Journal of Infection Control 787–792.

Schroeder, J, et al. "Handshaking promotes deal-making by signaling cooperative intent" (May 2019) 116(5) Journal of Personality and Social Psychology.

Sklansky, M, et al. "Banning the handshake from the health care setting" (25 Jun 2014) 311(24) JAMA 2477–2478.

Wooller, S. "People with a strong handshake are more intelligent: Study" (23 Apr 2018) New York Post.

Hug

Cohen, S, et al. "Does hugging provide stress-buffering social support? A study of susceptibility to upper respiratory infection and illness" (19 Dec 2015) 26(2) Psychological Science 135–147—"Hugging may effectively convey social support."

Forsell, LM, et al. "Meanings of hugging: From greeting behavior to touching implications" (Jan 2012) 1 Comprehensive Psychology.

Murphy, MLM, et al. "Receiving a hug is associated with the attenuation of negative mood that occurs on days with interpersonal conflict" (3 Oct 2018) 13(10) PLoS ONE.

Robinson, KJ, et al. "When in doubt, reach out: Touch is a covert but effective mode of soliciting and providing social support" (12 May 2015) 6(7) Social Psychological and Personality Science 831–839.

Shiomi, M, et al. "A hug from a robot encourages prosocial behaviour" (2017) 26th Institute of Electrical and Electronics Engineers International Symposium on Robot and Interactive

Communication 418–423—“Our experiment results with 38 participants showed that those who were hugged by a robot donated more money than those who only hugged the robot.”

Suvilehto, J, et al. “Topography of social touching depends on emotional bonds between humans” (26 Oct 2015) 112(45) Proceedings of the National Academy of Sciences—“These body regions formed relationship-specific maps in which the total area was directly related to the strength of the emotional bond between the participant and the touching person. Cultural influences were minor.”

Time panic!

Amabile, TM, et al. “Time pressure and creativity in organizations: A longitudinal field study” (Apr 2002) Harvard Business School Working Papers No. 01-073—“Time pressure on a given day negatively predicted creative cognitive processing that day, one day later, two days later, and over longer time periods as well.”

American Association for the Advancement of Science, “Research shows that busy people make healthier choices” (18 Sep 2018) EurekAlert—“When we perceive ourselves to be busy, it boosts our self-esteem, tipping the balance in favour of the more virtuous choice,” said Amitava Chattopadhyay, professor of marketing at INSEAD.

Bellezza, S, et al. “Conspicuous consumption of time: When busyness and lack of leisure time become a status symbol” (26 Dec 2017) 44(1) Journal of Consumer Research 118–138.

Bicknell, J. “Money doesn’t buy happiness—But time just might do it” (18 Jun 2018) Nautilus.

Burkeman, O. “Why you feel busy all the time (when you’re actually not)” (12 Sep 2016) BBC News.

“The case for a 4-day workweek?” (4 Sep 2018) Workforce Institute at Kronos Incorporated and Future Workplace—Survey of 3,000 employees. “Nearly half of employees worldwide could do their jobs in 5 hours or less each day.”

Cha, Y, et al. “Overwork and the slow convergence in the gender gap in wages” (8 Apr 2014) 79(3) American Sociological Review—“Because a greater proportion of men engage in overwork, these changes raised men’s wages relative to women’s and exacerbated the gender wage gap by an estimated 10 percent of the total wage gap.”

Chattopadhyay, A, et al. “Feel busy all the time? There’s an upside to that” (5 Jun 2018) Harvard Business Review

Collingwood, J. “Hofstadter’s Law and realistic planning” (8 Oct 2018) PsychCentral.

Curtin, M. “In an 8-Hour Day, the Average Worker Is Productive for This Many Hours” (21 July 2016) Inc.—“[T]he average worker is only productive for two hours and 53 minutes.”

Deloitte, “Meet the MilleXZials: Generational Lines Blur as Media Consumption for Gen X, Millennials and Gen Z Converge” (20 Mar 2018).

Dotti Sani, GM, et al. “Educational gradients in parents’ child-care time across countries,

1965–2012" (Apr 2016) 78(4) Journal of Marriage and Family—"In general, more educated mothers and fathers devoted more minutes to child care each day than less educated ones did."

Ebrahimi, M, et al. "To thrive or to suffer at the hand of busyness: How lay theories of busyness influence psychological empowerment and volunteering" (2017) 45 Advances in Consumer Research 79–84.

Etkin, J, et al. "Pressed for time? Goal conflict shapes how time is perceived, spent, and valued" (1 Jun 2015) 52(3) Journal of Marketing Research 394–406—"Encouraging consumers to take slow deep breaths or reappraise their anxiety as excitement can significantly reduce goal conflict's detrimental effects."

Festini, SB, et al. "The busier the better: Greater busyness is associated with better cognition" (17 May 2016) 8 Frontiers in Aging Neuroscience—"Although correlational, these data demonstrate that living a busy lifestyle is associated with better cognition."

Havas Group, "The modern nomad: Connect me if you can" (9 Sep 2015)—Survey of 10,131 men and women in 28 countries. "42% admit that they sometimes pretend to be busier than they actually are—and 6 in 10 believe that other people are faking their busyness."

Keinan, A, et al. "The symbolic value of time" (Apr 2019) 26 Current Opinion in Psychology 58–61—"Long hours of work and lack of leisure time have now become a status symbol."

Kim, JC, et al. "When busy is less indulging: Impact of busy mindset on self-control behaviors" (Feb 2019) 45(5) Journal of Consumer Research—"A busy mindset is predicted to facilitate people's ability to exert self-control."

Knecht, M, et al. "Going beyond work and family: A longitudinal study on the role of leisure in the work–life interplay" (4 Mar 2016) 37(7) Journal of Organizational Behaviour.

Levine, R. "Time use, happiness and implications for social policy: A report to the United Nations" (2013) 6(2) Insights—"Faster places were more economically healthy and residents tended to self-report being somewhat happier in their lives."

Livingston, G, et al. "7 facts about U.S. moms" (2018) Pew Research Center—"In 1965 women spent 10 hours a week on childcare. In 2016 that number increased to 14."

Locker, M. "Survey: Americans would pay $2,700 for an extra hour a day" (30 Oct 2014) Time. See also: BusinessWire, "Time is money: Cracking the code for balanced living."

MacKay, J. "Productivity in 2017: What we learned from analyzing 225 million hours of work time" (Jan 2018) RescueTime—"We only have 12.5 hours a week to do productive work."

Mark, G, et al. "No task left behind? Examining the nature of fragmented work" (2005) Proceedings of the 2005 CHI Conference on Human Factors in Computing Systems—Multitasking can lead to "stress in keeping track of multiple states of tasks."

Miller, C. "Women did everything right. Then work got greedy." (26 Apr 2019) New York Times—"Today, people who work 50 hours or more earn up to 8 percent more an hour than similar people working 35 to 49 hours."

Mogilner, C. "It's time for happiness" (19 Jul 2019) 26 Current Opinion in Psychology 80–84.

Newby-Clark, IR, et al. "People focus on optimistic scenarios and disregard pessimistic scenarios while predicting task completion times" (Sep 2000) 6(3) Journal of Experimental Psychology 171–182.

Oswald, AJ, et al. "Happiness and productivity" (7 Aug 2015) 33(4) Journal of Labor Economics—In an experimental study designed to make some workers happier, they found a "12% greater productivity." More generally, "lower happiness is systematically associated with lower productivity."

Pearson, H. "The lab that knows where your time really goes" (21 Oct 2015) 526(7574) Nature—"[P]eople who guess that they work 75-hour weeks, for example, can [overestimate how much they work] by more than 50%, and those of certain professions—teachers, lawyers, police officers—overestimate by more than 20%."

Roser, M. "Working hours" (2019) Our World in Data.

Rudd, J. "Long working days can cause heart problems, study says" (14 Jul 2017) Guardian.

Rudd, M. "Expand your breath, expand your time: Slow controlled breathing boosts time affluence" (2014) 42 Advances in Consumer Research 163–167.

Rudd, M. "Feeling short on time: Trends, consequences, and possible remedies" (Apr 2019) 26 Current Opinion in Psychology 5–10—Excellent review of time famine issue. "Myriad surveys have recently found that approximately two thirds of Americans say that they always or sometimes feel rushed and half say that they almost never feel that they have time on their hands."

Shepperd, JA, et al. "A primer on unrealistic optimism" (Jun 2015) 24(3) Current Directions in Psychological Science 232–237—"People in general are quite unrealistic in their estimates of the time it will take to complete a task, a misjudgment known as the *planning fallacy*."

Swant, M. "We're not nearly as busy as we pretend to be, according to a new study" (10 Sep 2015) Adweek—42 percent admit that they overstate how busy they are.

Weller, C. "Forget the 9 to 5—research suggests there's a case for the 3-hour workday" (26 Sep 2017) Business Insider.

Wepfer, AG, et al. "Work-life boundaries and well-being: Does work-to-life integration impair well-being through lack of recovery?" (Dec 2018) 33(6) Journal of Business and Psychology—"Employees who scored high on work-to-life integration enactment reported less recovery activities and in turn were more exhausted and experienced less work-life balance."

Whillans, AV, et al. "Buying time promotes happiness" (2017) Proceedings of the National Academy of Sciences—"Working adults report greater happiness after spending money on a time-saving purchase than on a material purchase." This research reveals a previously unexamined route from wealth to well-being: spending money to buy free time.

"Why is everyone so busy?" (20 Dec 2014) Economist—"The problem, then, is less how much time people have than how they see it."

Wilcox, K, et al. "How being busy can increase motivation and reduce task completion time" (Mar 2016) 110(3) Journal of Personality and Social Psychology 371–384—Study finds that people who feel busy are more likely to finish tasks and take less time to do so, probably because they are motivated to use their time efficiently.

Yang, AX, et al. "Idleness versus busyness" (Apr 2019) 26 Current Opinion in Psychology 15–18—"People pursue goals in order to engage in activities."

PART III: EVENING

Exercise

Alizadeh, Z, et al. "Comparison between the effect of 6 weeks of morning or evening aerobic exercise on appetite and anthropometric indices: A randomized controlled trial" (Jun 2017) 7(3) Clinical Obesity 157–165—"It appears that moderate- to high-intensity aerobic exercise in the morning could be considered a more effective programme than evening exercise on appetite control, calorie intake and weight loss in inactive overweight women."

Blackwell, D, et al. "State variation in meeting the 2008 federal guidelines for both aerobic and muscle-strengthening activities" (Jun 2018) 112 National Health Statistics Reports—"22.9 % of U.S. adults aged 18–64 met the guidelines for both aerobic and muscle-strengthening activities."

Brooker, P. "The feasibility and acceptability of morning versus evening exercise for overweight and obese adults: A randomized controlled trial" (11 Jan 2019) 14 Contemporary Clinical Trials Communications.

Burman, M, et al. "Does nighttime exercise really disturb sleep? Results from the 2013 National Sleep Foundation Sleep in America Poll" (Jul 2014) 15(7) Sleep Medicine—"Evening exercise was not associated with worse sleep. These findings add to the growing body of evidence that sleep hygiene recommendations should not discourage evening exercise."

Carlson, L. "Influence of exercise time of day on salivary melatonin responses" (1 Mar 2019) 14(3) Human Kinetics Journal—"If sleep is an issue, morning exercise may be preferable to afternoon exercise."

Cell Press, "Two studies explore whether time of day can affect the body's response to exercise" (18 Apr 2019) Medical Xpress—"Exercise in the evening seems to be more productive."

Chtourou, H, et al. "The effect of training at the same time of day and tapering period on the diurnal variation of short exercise performances" (Mar 2012) 26(3) Journal of Strength & Conditioning Research 697–708.

Colley, R, et al. "Comparison of self-reported and accelerometer-measured physical activity in Canadian adults" (19 Dec 2018) 29(12) Statistics Canada Health Reports—"On average, Canadian adults reported more physical activity than they accumulated on an accelerometer (49 minutes versus 23 minutes per day)."

Colley, R, et al. "Physical activity of Canadian children and youth, 2007 to 2015" (18 Oct

2017) 28(10) Statistics Canada Health Reports—"Data from the most recent cycle of the Canadian Health Measures Survey indicate that 7% of children and youth accumulated at least 60 minutes of MVPA [moderate-to-vigorous physical activity] on at least 6 out of 7 days."

Gordon, B, et al. "Afternoon but not morning exercise lowers blood glucose concentrations" (Jan 2017) 20 Journal of Science and Medicine in Sport.

Larsen, P, et al. "Evening high-intensity interval exercise does not disrupt sleep or alter energy intake despite changes in acylated ghrelin in middle-aged men" (29 Mar 2019) 104(6) Experimental Physiology—"HIIE [high-intensity interval exercise] can be performed in the early evening without subsequent sleep disruptions."

Statistics Canada, "Ten years of measuring physical activity—What have we learned?" (24 Nov 2017).

Stutz, J, et al. "Effects of evening exercise on sleep in healthy participants: A systematic review and meta-analysis" (Feb 2019) 49(2) Sports Medicine—"Overall, the studies reviewed here do not support the hypothesis that evening exercise negatively affects sleep, in fact rather the opposite."

Vitale, J, et al. "Sleep quality and high intensity interval training at two different times of day: A crossover study on the influence of the chronotype in male collegiate soccer players" (2017) 34(2) Chronobiology International.

Yamanaka, Y, et al. "Morning and evening physical exercise differentially regulate the autonomic nervous system during nocturnal sleep in humans" (1 Nov 2015) 309(9) American Journal of Physiology—Regulatory, Integrative and Comparative Physiology.

Youngstedt, S, et al. "Human circadian phase–response curves for exercise" (Apr 2019) 597(8) Journal of Physiology.

Hang with kids

Archer, C, et al. "Mother, baby and Facebook makes three: Does social media provide social support for new mothers?" (27 Jun 2018) 168(1) Media International Australia.

Blakemore, E. "It doesn't matter how much time parents spend with their kids" (30 Mar 2015) Smithsonian.

Borelli, JL, et al. "Bringing work home: Gender and parenting correlates of work-family guilt among parents of toddlers" (17 Mar 2017) 26 Journal of Child and Family Studies 1734–1745—"Mothers reported significantly higher levels of WIF-guilt [work-interfering-with family guilt] than fathers."

Chae, I. "'Am I a better mother than you?' Media and 21st-century motherhood in the context of the social comparison theory" (1 Jun 2015) 42(4) Communication Research 503–525.

Coyne, SM, et al. "'Do you dare to compare?' Associations between maternal social comparisons on social networking sites and parenting, mental health, and romantic relationship outcomes" (May 2017) 70(C) Computers in Human Behavior

335–340—"Results revealed that making social comparisons on social networking sites was related to parenting outcomes (in the form of higher levels of parental role overload, and lower levels of parental competence and perceived social support)."

Dotti Sani, GM, et al. "Educational gradients in parents' child-care time across countries, 1965–2012" (19 Apr 2016) 78(4) Journal of Marriage and the Family—"For both mothers and fathers, results indicated a widespread educational gradient and an increase in child-care time."

Farm Rich, "The 'guilty truth'—New research reveals top reasons for parental guilt" (13 Sep 2017) PR Newswire—"The new national survey of 2,000 parents of school-aged children, commissioned by Farm Rich," found that "American parents feel an average of 23 pangs of guilt per week."

Fomby, P, et al. "Mothers' time, the parenting package, and links to healthy child development" (26 Jul 2017) 80(1) Journal of Marriage and Family.

Gervis, Z. "Most parents think they're not making enough family memories" (15 Mar 2018) New York Post.

Hernández-Alava, M, et al. "Children's development and parental input: Evidence from the UK millennium cohort study" (Apr 2017) 54(2) Demography 485–511.

Hsin, A, et al. "When does time matter? Maternal employment, children's time with parents, and child development" (Oct 2014) 51(5) Demography 1867–1894—"On average, maternal work has no effect on time in activities that positively influence children's development, but it reduces time in types of activities that may be detrimental to children's development."

Hubert, S, et al. "Parental burnout: When exhausted mothers open up" (26 Jun 2018) 9 Frontiers in Psychology.

Kremer-Sadlik, T, et al. "Everyday moments: Finding 'quality time' in American working families" (2007) 16(2–3) Time and Society—"Everyday activities (like household chores or running errands) may afford families quality moments, unplanned, unstructured instances of social interaction that serve the important relationship-building functions that parents seek from 'quality time.'"

Logan, J, et al. "When children are not read to at home: The million word gap" (Jun 2019) 40(5) Journal of Developmental and Behavioral Pediatrics—"Parents who read 1 picture book with their children every day provide their children with exposure to an estimated 78,000 words each a year."

McGinn, K, et al. "Learning from Mum: Cross-national evidence linking maternal employment and adult children's outcomes" (30 Apr 2018) 33(3) Work, Employment and Society.

Mikel, B. "Harvard study: Kids of working moms grow up just as happy as stay-at-home moms" (21 Jul 2018) Inc.

Milkie, MA, et al. "Does the amount of time mothers spend with children or adolescents matter?" (Apr 2015) 77(2) Journal of Marriage and Family—"In childhood and adolescence, the amount of maternal time did not matter for offspring behaviors, emotions, or academics, whereas social status factors were important."

Milkie, MA, et al. "Time deficits with children: The link to parents' mental and physical health" (9 May 2018) Society and Mental Health—"It is unclear exactly what employed parents think is remiss, timewise. Ironically, many parents perceive time deficits, and this affects them, even though relative to earlier generations, they spend plenty of time with offspring."

Milkie, MA, et al. "What kind of war? 'Mommy Wars' discourse in U.S. and Canadian news, 1989–2013" (2016) 86(1) Sociological Inquiry 51–78.

Miller, CC. "Mounting evidence of advantages for children of working mothers" (15 May 2015) New York Times.

Opondo, C, et al. "Father involvement in early child-rearing and behavioural outcomes in their pre-adolescent children: Evidence from the ALSPAC UK birth cohort" (22 Nov 2016) 6(11) BMJ Open—"How new fathers see themselves as parents and adjust to the role, rather than the quantity of direct involvement in childcare, is associated with positive behavioural outcomes in children."

"Parents now spend twice as much time with their children as 50 years ago" (27 Nov 2017) Economist.

Roskam, I, et al. "Exhausted parents: Development and preliminary validation of the parental burnout inventory" (9 Feb 2017) 8 Frontiers in Psychology.

Schulte, B. "Does parent time matter for kids? Your questions answered" (1 Apr 2015) Washington Post.

Thompson, K. "How social media is making parenting more competitive than ever" (6 May 2016) Toronto Star.

Thomsen, MK. "Parental time investments in children: Evidence from Denmark" (27 Feb 2015) 58(3) Acta Sociologica 249—"The study initially found a positive and significant overall association between developmental care and children's educational performance."

"Today's parents spend more time with their kids than moms and dads did 50 years ago" (28 Sep 2016) UCI News.

Varathan, P. "Modern parents spend more time with their kids than their parents spent with them" (30 Nov 2017) Quartz.

Wolfers, J. "Yes, your time as a parent does make a difference" (1 Apr 2015) New York Times.

Check phone, again

Allred, RJ, et al. "The 'mere presence' hypothesis: Investigating the nonverbal effects of cell-phone presence on conversation satisfaction" (2017) 68(1) Communication Studies 22–36—"Whereas the mere presence of a cell phone did not influence conversation satisfaction, individuals' recollection of whether or not a cell phone was present did significantly negatively impact their pre- to posttest reports of conversation satisfaction."

Chotpitayasunondh, V, et al. "The effects of 'phubbing' on social interaction" (25 Mar 2018) Journal of Applied Social Psychology 48(6).

Crowley, JP, et al. “Replication of the mere presence hypothesis: The effects of cell phones on face-to-face conversations” (14 May 2018) 69(3) Communication Studies 283–293.

Davey, S, et al. “Predictors and consequences of ‘phubbing’ among adolescents and youth in India: An impact evaluation study” (Jan–Apr 2018) 25(1) Journal of Family and Community Medicine 35–42—“Phubbing also had significant consequences on their social health, relationship health, and self-flourishing, and was significantly related to depression and distress.”

Duke, K. “Cognitive costs of the mere presence of smartphones” (18 Aug 2017) Nature/NPJ Science of Learning—“Students performed worst on the cognitive tasks when their phones were on the desks in front of them, and best when their phones were in another room.”

Dwyer, RJ, et al. “Smartphone use undermines enjoyment of face-to-face social interactions” (Sep 2018) 78 Journal of Experimental Social Psychology 233–239—“In the real-world setting of a café, we found that people enjoyed a meal with their friends less when phones were present than when phones were put away. They also felt more distracted when phones were present (vs. absent), which had negative downstream consequences for their broader subjective experience (e.g., more tense arousal and boredom).”

Ha, TH. “The beginning of silent reading changed Westerners’ interior life” (19 Nov 2017) Quartz.

Han, S, et al. “Understanding nomophobia: Structural equation modeling and semantic network analysis of smartphone separation anxiety” (Jul 2017) 20(7) Cyberpsychology, Behavior, and Social Networking—“When users perceive smartphones as their extended selves, they are more likely to get attached to the devices, which, in turn, leads to nomophobia by heightening the phone proximity-seeking tendency.”

Hunter, JF, et al. “The use of smartphones as a digital security blanket: The influence of phone use and availability on psychological and physiological responses to social exclusion” (May 2018) 80(4) Psychosomatic Medicine 345–352—“The mere presence of a phone (and not necessarily phone use) can buffer against the negative experience and effects of social exclusion.”

Kushlev, K, et al. “Smartphones distract parents from cultivating feelings of connection when spending time with their children” (10 Apr 2018) 36(6) Journal of Social and Personal Relationships.

Kushlev, K, et al. “Smartphones reduce smiles between strangers” (2019) 91 Computers in Human Behavior 12–16—“Strangers smiled less to one another when they had their phones in a waiting room . . . These findings are based on objective behavioral coding rather than self-report and provide clear evidence that being constantly connected to the digital world may undermine important approach behavior.”

Lin, HL. “How your cell phone hurts your relationships” (4 Sep 2012) Scientific American.

Misra, S, et al. “The iPhone effect: The quality of in-person social interactions in the presence of mobile devices” (2016) 48(2) Environment and Behavior—“People who had conversations in the absence of mobile devices reported higher levels of empathetic concern."

"1 in 10 of us check our smartphones during sex—seriously" (13 May 2016) Telegraph—Reports on study notes 95% use phones in social situations.

Przybylski, AK, et al. "Can you connect with me now? How the presence of mobile communication technology influences face-to-face conversation quality" (19 Jul 2012) 30(3) Journal of Social and Personal Relationships—"The presence of mobile phones can interfere with human relationships, an effect that is most clear when individuals are discussing personally meaningful topics."

Roberts, JA, et al. "My life has become a major distraction from my cell phone: Partner phubbing and relationship satisfaction among romantic partners" (Jan 2016) 54 Computers in Human Behavior 134–141.

Tams, S, et al. "Smartphone withdrawal creates stress: A moderated mediation model of nomophobia, social threat, and phone withdrawal context" (Apr 2018) 81 Computers in Human Behavior—"The proposed indirect effect is nonsignificant only when situational certainty and controllability come together, that is, when people know for how long they will not be able to use their phones and when they have control over the situation."

Ward, AF, et al. "Brain drain: The mere presence of one's own smartphone reduces available cognitive capacity" (Apr 2017) 2(2) Journal of the Association for Consumer Research—"Even when people are successful at maintaining sustained attention—as when avoiding the temptation to check their phones—the mere presence of these devices reduces available cognitive capacity."

Wilmer, HH, et al. "Smartphones and cognition: A review of research exploring the links between mobile technology habits and cognitive functioning" (25 Apr 2017) 8 Frontiers in Psychology—"Empirical research on the cognitive impacts of smartphone technology is still quite limited."

Dinner

Danesi, G. "Pleasures and stress of eating alone and eating together among French and German young adults" (2012) School for Advanced Studies in the Social Sciences—Study involving German and French young adults and their perspectives regarding eating alone.

Dwyer, L, et al. "Promoting family meals: A review of existing interventions and opportunities for future research" (22 Jun 2015) 6 Adolescent Health, Medicine and Therapeutics 115–131—An analysis of various strategies to increase the frequency of family meals.

Ewa, J. "Class and eating: Family meals in Britain" (1 Sep 2017) 116 Appetite 527–535.

The Family Dinner Project, "Benefits of family dinners"—Useful resources from a project devoted to promoting family dinners.

Fishel, A. "Science says: Eat with your kids" (9 Jan 2015) The Conversation—Useful summary of the data.

Ghobadi, S, et al. "Association of eating while television viewing and overweight/obesity among children and adolescents: A systematic review and meta-analysis of observational studies" (Mar 2018) 19(3) Obesity Reviews—"Eating while TVV [television viewing] could be a risk factor for being overweight or obese in childhood and adolescents."

Gillman, M, et al. "Family dinner and diet quality among older children and adolescents" (2000) 9(3) Archives of Family Medicine 235–240—"Eating family dinner was associated with healthful dietary intake patterns, including more fruits and vegetables, less fried food and soda, less saturated and trans fat, lower glycemic load, more fiber and micronutrients from food, and no material differences in red meat or snack foods."

Hammons, AJ, et al. "Is frequency of shared family meals related to the nutritional health of children and adolescents?" (Jun 2011) 127(6) Pediatrics 1565–1574—"Children and adolescents who share family meals 3 or more times per week are more likely to be in a normal weight range and have healthier dietary and eating patterns than those who share fewer than 3 family meals together. In addition, they are less likely to engage in disordered eating."

Harbec, MJ, et al. "Associations between early family meal environment quality and later well-being in school-age children" (Feb–Mar 2018) 39(2) Journal of Developmental & Behavioral Pediatrics 136–143.

Harrison, ME, et al. "Systematic review of the effects of family meal frequency on psychosocial outcomes in youth" (Feb 2015) 61(2) Canadian Family Physician 96–106—"This systematic review provides further support that frequent family meals should be endorsed. All health care practitioners should educate families on the benefits of having regular meals together as a family."

Kwon, A, et al. "Eating alone and metabolic syndrome: A population-based Korean National Health and Nutrition Examination Survey 2013–2014" (Mar–Apr 2018) 12(2) Obesity Research & Clinical Practice 146–157—"Eating alone may be a potential risk factor for MetS [Metabolic syndrome]."

Litterbach, E. "Family meals with young children: An online study of family mealtime characteristics, among Australian families with children aged six months to six years" (24 Jan 2017) 17 BioMed Central Public Health.

Livingstone, M, et al. "Portion size and obesity" (3 Nov 2014) 5(6) Advances in Nutrition 829–834.

Mills, S, et al. "Frequency of eating home cooked meals and potential benefits for diet and health: Cross-sectional analysis of a population-based cohort study" (17 Aug 2017) 14(1) International Journal of Behavioral Nutrition and Physical Activity—"Eating home cooked meals more frequently was associated with better dietary quality and lower adiposity."

Robinson, E, et al. "Portion size and later food intake: Evidence on the 'normalizing' effect of reducing food portion sizes" (1 Apr 2018) 107(4) American Journal of Clinical Nutrition 640–646.

Takeda, W, et al. "Spatial, temporal, and health associations of eating alone: A cross-cultural analysis of young adults in urban Australia and Japan" (2017) 118 Appetite 149–160—Interesting study that explored the cultural associations with eating alone. Some Australians, for example, associated it with healthier eating.

Tani, Y, et al. "Combined effects of eating alone and living alone on unhealthy dietary behaviors, obesity and underweight in older Japanese adults" (Dec 2015) 95 Appetite 1–8—"Eating alone was associated with unhealthy dietary behaviors in older men and women."

Utter, J, et al. "Feasibility of a family meal intervention to address nutritional wellbeing, emotional wellbeing and food insecurity of families with adolescents" (Jul–Aug 2018) 50(7) Journal of Nutrition Education and Behavior—"Providing families with meal plans, recipes and ingredients is an acceptable way to improve family meals."

Vik, FN, et al. "Associations between eating meals, watching TV while eating meals and weight status among children, ages 10–12 years in eight European countries: The ENERGY cross-sectional study" (13 May 2013) 10 International Journal of Behavioral Nutrition and Physical Activity.

Walton, K, et al. "Exploring the role of family functioning in the association between frequency of family dinners and dietary intake among adolescents and young adults" (2 Nov 2018) 1(7) JAMA Network Open—"More frequent family dinners are associated with healthful dietary intakes among youths, regardless of level of family functioning. Family dinners may be an appropriate intervention target for improving dietary intake among youths."

Wolfson, JA, et al. "Is cooking at home associated with better diet quality or weight-loss intention?" (Jun 2015) 18(8) Public Health Nutrition 1397–1406—"Cooking dinner frequently at home is associated with consumption of a healthier diet whether or not one is trying to lose weight."

Ziauddeen, N, et al. "Eating at food outlets and leisure places and 'on the go' is associated with less-healthy food choices than eating at home and in school in children" (1 Jun 2018) 107(6) American Journal of Clinical Nutrition 992–1003—"Home and school eating are associated with better food choices, whereas other locations are associated with poor food choices."

Wine

Almenberg, J, et al. "When does the price affect the taste? Results from a wine experiment" (Jan 2011) 6(1) Journal of Wine Economics 111–121—"Disclosing a high price before tasting the wine produces considerably higher ratings."

Berns, GS. "Price, placebo, and the brain" (1 Nov 2005) 42(4) Journal of Marketing Research—"Through repeated exposure, higher-priced items tend to be associated with better quality goods and services and, therefore, are expected to deliver more utility to a consumer."

Bohannon, J, et al. "Can people distinguish pâté from dog food?" (Apr 2009) American Association of Wine Economists Working Papers 36—"Subjects were not better than random at correctly identifying the dog food."

Centers for Disease Control and Prevention, "Moderate drinking" (18 Oct 2016).

Danner, L, et al. "Context and wine quality effects on consumers' mood, emotions, liking and willingness to pay for Australian Shiraz wines" (Nov 2016) 89(1) Food Research International 254–265—Context matters to taste.

Danner, L, et al. "'I like the sound of that!' Wine descriptions influence consumers' expectations, liking, emotions and willingness to pay for Australian white wines" (Sep 2017) 99(1) Food Research International 263–274—"Elaborate information resulted in highest liking, WTP [willingness-to-pay] and positive emotions . . . and a substantial increase in willingness to pay after tasting."

Doucleff, M. "Drinking with your eyes: How wine labels trick us into buying" (11 Oct 2013) NPR.

Enax, L, et al. "Marketing placebo effects—From behavioral effects to behavior change?" (Nov 2015) 13(1) Journal of Agricultural & Food Industrial Organization 15–31.

Goldstein, R, et al. "Do more expensive wines taste better? Evidence from a large sample of blind tastings" (Spring 2008) 3(1) Journal of Wine Economics—"In a large sample of blind tastings, we find that the correlation between price and overall rating is small and negative. Unless they are experts, individuals on average enjoy more expensive wines slightly less."

Haseeb, S, et al. "Wine and cardiovascular health: A comprehensive review" (10 Oct 2017) 136(15) Circulation—"Although there is extensive epidemiological support for this drinking pattern [light-to-moderate intake], a consensus has not been reached."

Hodgson, RT. "An examination of judge reliability at a major U.S. wine competition" (Winter 2008) 3(2) Journal of Wine Economics 105–113.

Lee, WF, et al. "Effect of extrinsic cues on willingness to pay [WTP] of wine" (5 Nov 2018) 120(11) British Food Journal.

McLaughlin, R, et al. "Putting coffee to the test: Does pricier java really taste better?" (5 Feb 2018) CTV News Vancouver.

Morrot, G, et al. "The color of odors" (Nov 2001) 79(2) Brain and Language—The famous study that found experts couldn't tell the difference between red and white wine.

Parr, WV. "Demystifying wine tasting: Cognitive psychology's contribution" (Oct 2019) 124 Food Research International.

Piqueras-Fiszman, B, et al. "Sensory expectations based on product-extrinsic food cues: An interdisciplinary review of the empirical evidence and theoretical accounts" (2015) 40(A) Food Quality and Preference 165–179.

Plassmann, H, et al. "Marketing actions can modulate neural representations of experienced pleasantness" (Jan 2008) 105(3) Proceedings of the National Academy of Sciences—"Increasing the price of a wine increases subjective reports of flavor pleasantness as well as blood-oxygen-level-dependent activity in medial orbitofrontal cortex."

Pomeroy, R. "The legendary study that embarrassed wine experts across the globe" (18 Aug 2014) RealClearScience.

Sample, I. "Expensive wine and cheap plonk taste the same to most people" (14 Apr 2011)

Guardian—Report on the study that found people could not reliably identify cheap and expensive wine. "People just could not tell the difference between cheap and expensive wine . . . When you know the answer, you fool yourself into thinking you would be able to tell the difference, but most people simply can't."

Schmidt, L. "How context alters value: The brain's valuation and affective regulation system link price cues to experienced taste pleasantness" (14 Aug 2017) 7 Scientific Reports.

Shiv, B, et al. "Placebo effects of marketing actions: Consumer may get what they pay for" (Nov 2005) 42(4) Journal of Marketing Research—"Consumers who pay a discounted price for a product (e.g., an energy drink thought to increase mental acuity) may derive less actual benefit from consuming this product."

Siegrist, M, et al. "Expectations influence sensory experience in a wine tasting" (Jun 2009) 52(3) Appetite 762—"When the information was given prior to the tasting, negative information about the wine resulted in lower ratings compared to the group that received positive information."

Stanley, TL. "Payless opened a fake luxury store, 'Palessi,' to see how much people would pay for $20 shoes" (28 Nov 2018) Adweek.

Swerdloff, A. "The majority of people might just genuinely prefer cheap coffee" (26 Aug 2016) Vice.

University of Bonn, "Why expensive wine appears to taste better" (14 Aug 2017) Phys.org.

Wang, Q, et al. "Assessing the influence of music on wine perception among wine professionals" (Mar 2018) 6(2) Food Science & Nutrition 295–301.

Wang, QJ, et al. "Does blind wine tasting work? Investigating the impact of training on blind tasting accuracy and wine preference" (2018) Science & Wine.

Wood, A, et al. "Risk thresholds for alcohol consumption: Combined analysis of individual-participant data for 599 912 current drinkers in 83 prospective studies" (14 Apr 2018) 391(10129) The Lancet.

World Health Organization, "Alcohol" (21 Sep 2018).

Zeidler, M. "Like the label? You'll probably like the wine, says UBC researcher" (17 Mar 2019) CBC News.

Dishes

Carlson, DL, et al. "The gendered division of housework and couples' sexual relationships: A reexamination" (25 May 2016) 78(4) Journal of Marriage and Family.

Carlson, DL, et al. "Sharing's more fun for everyone? Gender attitudes, sexual self-efficacy, and sexual frequency" (Aug 2018) 81(3) Journal of Marriage and Family.

Carlson, DL, et al. "Stalled for whom? Change in the division of particular housework tasks and their consequences for middle- to low-income couples" (6 Apr 2018) 4 Socius: Sociological Research for a Dynamic World 1–17—"The division of dishwashing, among all tasks, is most consequential to relationship quality, especially for women."

Council on Contemporary Families, "Not all housework is created equal: Particular housework tasks and couples' relationship quality" (3 Apr 2018).

Johnson, MD, et al. "Skip the dishes? Not so fast! Sex and housework revisited" (Mar 2016) 30(2) Journal of Family Psychology—"This study provides a robust counterpoint to recent findings suggesting that men's participation in housework is harmful to a couple's sex life."

Toilet seat up or down?

Case, MA. "Why not abolish the laws of urinary segregation?" in Molotch, H, et al. (eds) *Toilet: Public Restrooms and the Politics of Sharing* (New York University Press, 2010).

Choi, JP. "Up or down? A male economist's manifesto on the toilet seat etiquette" (Nov 2002) Michigan State University Working Papers.

General, R. "33% of Japanese men in survey prefer sitting down while peeing" (11 Sep 2018) NextShark.

Moss, G. "7 reasons dudes should be the ones to put the toilet seat back down—Every. Single. Time." (20 Mar 2015) Bustle.

Nonaka, R. "44% of men pee at home while sitting down, survey reveals" (Dec 2017) Asahi Shimbun.

Siddiqi, H. "The social norm of leaving the toilet seat down: A game theoretic analysis" (29 May 2007) Science Creative Quarterly.

Stamp, J. "From turrets to toilets: A partial history of the throne room" (20 Jun 2014) Smithsonian.

Wells, J. "Should men put the toilet seat down when they're finished?" (11 Oct 2015) Telegraph.

10,000 steps?

Cox, D. "Watch your step: Why the 10,000 daily goal is built on bad science" (3 Sep 2018) Guardian.

Cummins, E. "24/7 fitness trackers won't solve all your problems—and they might make you imagine new ones" (4 Mar 2019) Popular Science.

Duke University, "Why counting your steps could make you unhappier" (21 Dec 2015) Fuqua School of Business.

Etkins, J. "The hidden cost of personal quantification" (Apr 2016) 42(6) Journal of Consumer Research—Activity tracking can make exercise less enjoyable.

Feehan, L, et al. "Accuracy of Fitbit devices: Systematic review and narrative syntheses of quantitative data" (9 Aug 2018) 6(8) JMIR Health—"Other than for measures of steps in adults with no limitations in mobility, discretion should be used when considering the use of Fitbit devices as an outcome measurement tool in research or to inform health care decisions, as there are seemingly a limited number of situations where the device is likely to provide accurate measurement."

Finkelstein, E. "Effectiveness of activity trackers with and without incentives to increase physical activity (TRIPPA): A randomised controlled trial" (1 Dec 2016) 4(12) The Lancet Diabetes & Endocrinology—"We identified no evidence of improvements in health outcomes, either with or without incentives, calling into question the value of these devices for health promotion."

"Fitness trackers 'overestimate' calorie burning" (28 Jan 2019) BBC News.

Heathman, A. "Your fitness tracker is probably overestimating the calories you're burning" (28 Jan 2019) Evening Standard.

Jakicic, J, et al. "Effect of wearable technology combined with a lifestyle intervention on long-term weight loss" (20 Sep 2016) 316(11) JAMA—"Devices that monitor and provide feedback on physical activity may not offer an advantage over standard behavioral weight loss approaches."

James, T, et al. "Using organismic integration theory to explore the associations between users' exercise motivations and fitness technology feature set use" (Mar 2019) 43(1) MIS Quarterly—"The social interaction and data management features of current fitness technologies show promise in assisting well-being outcomes, but only for the more self-determined and amotivated subtypes of exercisers."

Kerner, C. "The motivational impact of wearable healthy lifestyle technologies: A self-determination perspective on Fitbits with adolescents" (12 Apr 2017) 48(5) American Journal of Health Education—"Findings suggest that healthy lifestyle technology may have negative motivational consequences."

Thosar, S, et al. "Self-regulated use of a wearable activity sensor is not associated with improvements in physical activity, cardiometabolic risk or subjective health status" (Sep 2018) 52(18) British Journal of Sports Medicine—Study found that activity actually decreased with use, though participants thought they were more active.

Binge-watch TV

Ahmed, AAM. "New era of TV-watching behavior: Binge watching and its psychological effects" (Jan 2017) 8(2) Media Watch 192–207.

American Academy of Sleep Medicine, "Sleep or Netflix? You can have both when you binge-watch responsibly" (30 May 2017).

Cakebread, C, "Here are all the reasons why Americans say they binge-watch TV shows" (15 Sep 2017) Business Insider.

Chambliss, C, et al. "Distracted by binge-watching: Sources of academic and social disruption in students" (2017) 3(1) Atlantic Research Centre Journal of Pediatrics—"A majority of college students (64%) reported excessive binge watching of Netflix and other non-sport TV programs."

De Feijter, D, et al. "Confessions of a 'guilty' couch potato understanding and using context to optimize binge-watching behaviour" (17 Jun 2016) Proceedings of the ACM

International Conference on Interactive Experiences for TV and Online Video—"An in-situ, smartphone monitoring survey among Dutch binge-watchers was used to reveal context factors related to binge-watching and wellbeing. Results indicate that binge-watching is a solitary activity that occurs in an online socially active context."

Deloitte, "Meet the MilleXZials: Generational Lines Blur as Media Consumption for Gen X, Millennials and Gen Z Converge" (20 Mar 2018).

Devasagayam, R. "Media bingeing: A qualitative study of psychological influences" (Mar 2014) Proceedings of the Marketing Management Association Annual Conference—"The results propose the forming of one-sided, unconscious bonds between viewers and characters. We believe this bond is one of the main factors influencing bingeing behaviors."

Exelman, L, et al. "Binge viewing, sleep, and the role of pre-sleep arousal" (15 Aug 2017) 13(8) Journal of Clinical Sleep Medicine 1001–1008—"Higher binge viewing frequency was associated with a poorer sleep quality, increased fatigue and more symptoms of insomnia, whereas regular television viewing was not."

Fine, D. "Fear not, technology isn't actually making us dumber" (21 Dec 2016) Sydney Morning Herald—Includes comment by Conrad Gessner.

Flayelle, M, et al. "Time for a plot twist: Beyond confirmatory approaches to binge-watching research" (Jan 2019) 8(3) Psychology of Popular Media Culture—This article has a nice review of the literature, particularly in the context of possible harms. It notes that "binge-watching remains an understudied phenomenon, despite its widespread manifestation in today's society."

Flayelle, M, et al. "Toward a qualitative understanding of binge-watching behaviors: A focus group approach" (1 Dec 2017) 6(4) Journal of Behavioral Addictions 457–471—"Undoubtedly, TV series watching, like any hobby or leisure activity, primarily satisfies the need for entertainment."

Grace, M, et al. "Television viewing time and inflammatory-related mortality" (Oct 2017) 49(10) Medicine & Science in Sports & Exercise 2040–2047—"Before adjustment for leisure-time physical activity, TV time was associated with increased risk of inflammatory-related mortality."

Horvath, JC, et al. "The impact of binge watching on memory and perceived comprehension." (4 Sep 2017) 22(9) First Monday—"These memories [of the TV shows] decay more rapidly than memories formed after daily- or weekly-episode viewing schedules. In addition, participants in the binge watching condition reported significantly less show enjoyment than participants in the daily- or weekly-viewing conditions."

Kubota, Y, et al. "TV viewing and incident venous thromboembolism: The Atherosclerotic Risk in Communities Study" (Apr 2018) 45(4) Journal of Thrombosis and Thrombolysis 353–359—"Greater frequency of TV viewing was independently associated with increased risk of VTE [venous thromboembolism], partially mediated by obesity. Achieving a recommended physical activity level did not eliminate the increased VTE risk associated with frequent TV viewing. Avoiding frequent TV viewing as well as increasing physical activity and controlling body weight might be beneficial for VTE prevention."

Morris, C. "Depression, disease and no sex are some dangers of binge watching" (26 Aug 2016) Consumer News and Business Channel.

Netflix, "Ready, set, binge: More than 8 million viewers 'binge race' their favorite series" (17 Oct 2017)—"In total, 8.4 million members have chosen to Binge Race during their Netflix tenure."

Olson, S. "Binge watching TV linked to higher rates of depression and anxiety" (8 Nov 2015) Medical Daily—Report on study that found that "after watching just two hours of TV participants reported feeling more depressed and anxious than those who spent fewer time watching TV."

Page, D. "What happens to your brain when you binge-watch a TV series" (4 Nov 2017).

Patient.info, "Over 50% of Brits suffer from post binge-watching blues, Patient.info reports" (28 Feb 2018) PR Newswire—"More than half of the 2,000 survey respondents admitted to having experienced mental health issues brought on by the end of a TV series."

Rigby, JM, et al. "'I can watch what I want': A diary study of on-demand and cross-device viewing" (26–28 Jun 2018) Proceedings of the ACM International Conference on Interactive Experiences for TV and Online Video—"Evening prime time continued to be the most popular time for people to watch on-demand content." "In total, 135 sessions (75.8%) were watched alone."

Rodriguez, A. "The average young American binge-watches TV for five hours straight" (23 Mar 2017) Quartz.

Spangler, T. "Binge nation: 70% of Americans engage in marathon TV viewing" (16 Mar 2016) Variety.

Spruance, LA, et al. "Are you still watching?: Correlations between binge TV watching, diet and physical activity" (14 Jul 2017) Journal of Obesity & Weight Management.

Sung, YH, et al. "A bad habit for your health? An exploration of psychological factors for binge watching behavior" (21 May 2015) 65th Annual Conference of the International Communication Association.

Sung, YH, et al. "Why Do We Indulge? Exploring Motivations for Binge Watching" (12 Jul 2018) 62(3) Journal of Broadcasting & Electronic Media 408—"Only the entertainment motivation is a significant predictor of binge watching for those with a low level of binge watching."

Tuck, "Streaming content and sleep—2018 study" (2 Aug 2018)—"Nearly half (45%) of adults have pulled an all-nighter in the last year to watch a TV show."

Walton-Pattison, E, et al. "'Just one more episode': Frequency and theoretical correlates of television binge watching" (2018) 23(1) Journal of Health Psychology.

Wash hair

"ASA finds TRESemme Naturals ads misleading" (5 Jul 2011) Cosmetic Business.

Brueck, H. "How often you actually need to shower, according to science" (1 Feb 2019) MSN.

"Claims in shampoo ad 'misleading'" (11 May 2005) BBC News.

Cruz, CF, et al. "Human hair and the impact of cosmetic procedures: A review on cleansing and shape-modulating cosmetics" (Jul 2016) 3(3) Cosmetics.

Dawber, R. "Hair: Its structure and response to cosmetic reparation" (1996) 14(1) Clinics in Dermatology 105–112.

De Blasio, B, et al. "From cradle to cane: The cost of being a female consumer" (Dec 2015) NYC Department of Consumer Affairs.

Draelos, ZD. "Essentials of hair care often neglected: Hair cleansing" (2010) 2(1) International Journal of Trichology 24–29—"Technically, it is not necessary to shampoo the hair daily unless sebum production is high. Shampooing is actually more damaging to the hair shaft than beneficial."

Elgart, O. "Revealed: How shampoo ads have us all fooled" (6 Apr 2018) New Zealand Herald—"69 per cent believe that haircare advertising is misleading."

"Facial moisturizers more expensive for women than for men" (1 May 2019) United Press International—"On average, the products marketed to women cost $3.09 more per ounce than those marketed to men."

Gray, J. "Hair care and hair care products" (Mar–Apr 2001) 19(2) Clinics in Dermatology 227–236.

Haskin, A, et al. "Breaking the cycle of hair breakage: Pearls for the management of acquired trichorrhexis nodosa" (Jun 2017) 28(4) Journal of Dermatological Treatment 322–326.

Kenneth, JA. "Rolling back the 'pink tax': Dim prospects for eliminating gender-based price discrimination in the sale of consumer goods and services" (2018) 54(2) California Western Law Review—"As the research overwhelmingly demonstrates, gender-based pricing—also known as the 'pink tax' or 'gender tax'—is a reality that cannot be explained other than by discrimination based solely on gender."

Khazan, O. "How often people in various countries shower" (17 Feb 2015) Atlantic.

Morales, T. "Are expensive shampoos better?" (11 Apr 2005) CBS News—"There is absolutely no difference between expensive products and inexpensive products, and I say that unequivocally."

Schlossberg, M. "30 items that prove women pay more than men for the same products" (16 Jul 2016) Business Insider.

Shaw, H. "'Pink tax' has women paying 43% more for their toiletries than men" (25 Apr 2016) Financial Post.

Trüeb, RM. "Shampoos: Ingredients, efficacy and adverse effects" (May 2007) 5(5) Journal der Deutschen Dermatologischen Gesellschaft 356–365.

Waters, L. "Does the price of your shampoo affect how clean your hair is? Here's the science" (23 Jan 2017) The Conversation.

Floss teeth

American Academy of Periodontology, "More than a quarter of U.S. adults are dishonest with dentists about how often they floss their teeth" (23 Jun 2015)—"More than a quarter (27%) of U.S. adults admit they lie to their dentist about how often they floss their teeth. Additionally, more than one-third of Americans (36%) would rather do an unpleasant activity like cleaning the toilet."

American Dental Association, "The medical benefit of daily flossing called into question" (2 Aug 2016)—"While the average benefit is small and the quality of the evidence is very low (meaning the true average benefit could be higher or lower), given that periodontal disease is estimated to affect half of all Americans, even a small benefit may be helpful."

American Dental Association, "New survey highlights 'unusual' flossing habits" (20 Oct 2017)—"16% said they always floss at least once a day . . . Forty-four percent of those surveyed admit they have exaggerated to their dentist about how much they floss when asked."

Cepeda, MS, et al. "Association of flossing/inter-dental cleaning and periodontitis in adults" (Sep 2017) 44(9) Journal of Clinical Periodontology 866–871—This study, which could not establish causation, found that "flossing was associated with a modestly lower prevalence of periodontitis . . . Flossing 2-4 days a week could be as beneficial as flossing more frequently."

De Oliveira, KMH, et al. "Dental flossing and proximal caries in the primary dentition: A systematic review" (2017) 15(5) Oral Health and Preventive Dentistry 427–434—"There is only one study in the current literature showing evidence of an association between the use of dental floss and proximal caries reduction on primary dentition."

"Dentists—Canada market research report" (Aug 2018) IBIS World—"Over the five years to 2019, the Dentists industry in Canada has exhibited growth due to an uptick in the use of dental services and increased expenditure on dental services."

Donn, J. "Medical benefits of dental floss unproven" (2 Aug 2016) Associated Press—"In a letter to the AP, the government acknowledged the effectiveness of flossing had never been researched, as required."

Fleming, EB, et al. "Prevalence of daily flossing among adults by selected risk factors for periodontal disease—United States, 2011–2014" (Aug 2018) 89(8) Journal of Periodontology 933–939—"Daily flossing was higher among women, those with higher income." Also, 42.2 percent of tobacco users never floss compared with 29.8 percent of those who don't use tobacco.

[illegible] "[illegible] of Americans lie to dentists about flossing" (Jul 2015) Scientific American—"Flossing one's teeth, according to a Harris Poll survey, is in some cases a less desirable activity than listening to the sound of nails on a chalkboard or to small children crying on a bus or plane."

Hamilton, K, et al. "Dental flossing and automaticity: A longitudinal moderated mediation analysis" (Jun 2018) 23(5) Psychology, Health & Medicine 619–627.

Hujoel, PP, et al. "Dental flossing and interproximal caries: A systematic review" (Apr 2006) 85(4) Journal of Dental Research 298–305—"Professional flossing performed on school days for 1.7 years on predominantly primary teeth in children was associated with a 40% caries risk reduction."

Hujoel, PP, et al. "Personal oral hygiene and dental caries: A systematic review of randomised controlled trials" (Dec 2018) 35(4) Gerodontology 282–289—"Personal oral hygiene in the absence of fluorides has failed to show a benefit in terms of reducing the incidence of dental caries."

Jupes, O. "Dentists have stopped being strung along by the great flossing yarn. About time" (3 Aug 2016) Guardian.

Kassebaum, NJ, et al. "Global burden of untreated caries: A systematic review and metaregression" (May 2015) 94(5) Journal of Dental Research 650–658—"The global age-standardized prevalence and incidence of untreated caries remained static between 1990 and 2010."

Knapton, S. "Flossing teeth does little good, investigation finds as US removes recommendation from health advice" (2 Aug 2016) Telegraph.

Kuru, BE, et al. "Role of the mechanical interdental plaque control in the management of periodontal health: How many options do we have?" in Gingival Disease—A Comprehensive and Professional Approach for Treatment and Prevention (5 Nov 2018)—"Current literature unfortunately does not support dental floss usage on a routine basis."

Lee, JH, et al. "Association of toothbrushing and proximal cleaning with periodontal health among Korean adults: Results from Korea National Health and Nutrition Examination Survey in year 2010 and 2012" (Mar 2018) 45(3) Journal of Periodontology 322–335.

Marchesan, JT, et al. "Interdental cleaning is associated with decreased oral disease prevalence" (Jul 2018) 97(7) Journal of Dental Research—An association was found between interdental cleaning and oral health, but only a correlation.

Mazhari, F, et al. "The effect of toothbrushing and flossing sequence on interdental plaque reduction and fluoride retention: A randomized controlled clinical trial" (Jul 2018) 89(7) Journal of Periodontology—"Flossing followed by brushing is preferred to brushing then flossing in order to reduce interdental plaque and increase fluoride concentration in interdental plaque."

Niederman, R. "Psychological approaches may improve oral hygiene behaviour" (25 Jun 2007) 8(2) Journal of Evidence-Based Dental Practice 39–40.

Ontario Dental Association, "Your oral health" (2017)—"In our clinical practice, we see evidence [that flossing works] every day. We find reduced incidence of tooth decay and healthier gum tissue in our patients who use floss or other methods to remove food debris and plaque between teeth."

Ritchey, G. "May the Floss be with you?" (6 Nov 2015) Science Based Medicine.

Sälzer, S, et al. "Efficacy of inter-dental mechanical plaque control in managing gingivitis—a meta-review" (Apr 2015) 42 Journal of Clinical Periodontology—"The majority of available studies fail to demonstrate that flossing is generally effective in plaque removal."

Sambunjak, D, et al. “Flossing for the management of periodontal diseases and dental caries in adults” (7 Dec 2011) 12(12) Cochrane Library—“There is some evidence from twelve studies that flossing in addition to toothbrushing reduces gingivitis compared to toothbrushing alone. There is weak, very unreliable evidence from 10 studies that flossing plus toothbrushing may be associated with a small reduction in plaque at 1 and 3 months. No studies reported the effectiveness of flossing plus toothbrushing for preventing dental caries.” (Note: this review was updated in April 2019, though the conclusions are similar. See Worthington H, et al. “Home use of interdental cleaning devices, in addition to toothbrushing, for preventing and controlling periodontal diseases and dental caries” (2019) 4 Cochrane Database of Systematic Reviews—“Overall, the evidence was low to very low-certainty, and the effect sizes observed may not be clinically important.”)

Vernon, LT, et al. “In defense of flossing: Can we agree it’s premature to claim flossing is ineffective to prevent dental caries?” (Jun 2017) 17(2) Journal of Evidence-Based Dental Practice 71–75.

Wilder, RS, et al. “Improving periodontal outcomes: Merging clinical and behavioral science” (Jun 2016) 71(1) Periodontology 2000 65–81—Nice review of the literature.

Sex

Anderson, RM. “Positive sexuality and its impact on overall well-being” (Feb 2013) 56(2) Bundesgesundheitsblatt–Gesundheitsforschung–Gesundheitsschutz—“Sexual health, physical health, mental health, and overall well-being are all positively associated with sexual satisfaction, sexual self-esteem, and sexual pleasure.”

Loewenstein, G, et al. “Does increased sexual frequency enhance happiness?” (Aug 2015) 116 Journal of Economic Behavior & Organization—“Increased frequency does not lead to increased happiness, perhaps because it leads to a decline in wanting for, and enjoyment of, sex.”

Muise, A, et al. “Sexual frequency predicts greater well-being, but more is not always better” (Nov 2015) 7(4) Social Psychology and Personality Science—“For people in relationships, sexual frequency is no longer significantly associated with well-being at a frequency greater than once a week.”

Smith, A, et al. “Sexual and relationship satisfaction among heterosexual men and women: The importance of desired frequency of sex” (2011) 37(2) Journal of Sex and Marital Therapy—“Only 46% of men and 58% of women were satisfied with their current frequency of sex.”

Strapagiel, L. “People think everyone is having a lot of sex, but a survey shows that’s not the case” (9 Aug 2018) BuzzFeed—“Men guessed that women had sex 23 times [within the last four weeks], but the actual number was an average of 5 to 6 times.”

Wadsworth, T. “Sex and the pursuit of happiness: How other people’s sex lives are related to our sense of well-being” (Mar 2014) 116(1) Social Indicators Research—“The overall process by which sex is associated with happiness is intricately connected to our perceptions of the sex lives of others."

Cuddle

Muise, A, et al. "Post sex affectionate exchanges promote sexual and relationship satisfaction" (Oct 2014) 43(7) Archives of Sexual Behavior 1391–1402—"The period after sex is a critical time for promoting satisfaction in intimate bonds."

Sleep

Baron, KG, et al. "Orthosomnia: Are some patients taking the quantified self too far?" (15 Feb 2017) 13(12) Journal of Clinical Sleep Medicine 351–354—"Most consumers are unaware that the claims of these devices often outweigh the science to support them as devices to measure and improve sleep."

Copland, S. "The many reasons that people are having less sex" (9 May 2017) BBC News.

Department of Health and Human Services, "Your guide to healthy sleep" (Sep 2011).

Division of Sleep Medicine at Harvard Medical School, "Consequences of Insufficient Sleep" Healthy Sleep—"Data from three large cross-sectional epidemiological studies reveal that sleeping five hours or less per night increased mortality risk from all causes by roughly 15 percent."

Duncan, MJ, et al. "Greater bed- and wake-time variability is associated with less healthy lifestyle behaviors: A cross-sectional study" (Feb 2016) 24(1) Journal of Public Health 31–40—"Having bedtimes that varied by >30 min were associated with lower dietary quality, higher alcohol consumption, higher sitting time, more frequent insufficient sleep and poorer overall pattern of lifestyle behaviors."

Feehan, LM, et al. "Accuracy of Fitbit devices: Systematic review and narrative syntheses of quantitative data" (9 Aug 2018) 6(8) JMIR mHealth and uHhealth.

Gervis, Z. "Phones turn bedrooms into a no-sex zone" (15 Aug 2018) New York Post.

Grandner, MA, et al. "Mortality associated with short sleep duration: The evidence, the possible mechanisms, and the future" (Jun 2010) 14(3) Sleep Medicine Reviews 191–203.

Hakim, M, et al. "Comparison of the Fitbit® Charge and polysomnography [PSG] for measuring sleep quality in children with sleep disordered breathing" (7 Nov 2018) Minerva Pediatrica—"The current prospective study confirms that the Fitbit® Charge overestimates time spent asleep compared to PSG in children with OSA/SDB symptoms, limiting the validity of sleep monitoring with wearable activity trackers appears in these patients."

Hughes, N, et al. "Sleeping with the frenemy: How restricting 'bedroom use' of smartphones impacts happiness and wellbeing" (Aug 2018) 85 Computers in Human Behavior.

Knapton, S. "Britons are having less sex, and Game of Thrones could be to blame" (5 Jun 2016) Telegraph.

Ko, PR, et al. "Consumer sleep technologies: A review of the landscape" (15 Dec 2015) 11(12) Journal of Clinical Sleep Medicine 1455–1461.

Lawrenson, J, et al. "The effect of blue-light blocking spectacle lenses on visual performance, macular health and the sleep-wake cycle: A systematic review of the literature" (Nov 2017) 37(6) Ophthalmic and Physiological Optics—"We find a lack of high quality evidence to support using BB [blue-light blocking] spectacle lenses for the general population to improve visual performance or sleep quality."

Lee, JM. "Comparison of wearable trackers' ability to estimate sleep" (15 Jun 2018) 15(6) International Journal of Environmental Research and Public Health.

Liang, Z, et al. "Validity of consumer activity wristbands and wearable EEG for measuring overall sleep parameters and sleep structure in free-living conditions" (Jun 2018) 2(1–2) Journal of Healthcare Informatics Research 152–178.

Lichstein, KL. "Insomnia identity" (Oct 2017) 97 Behaviour Research and Therapy 230–241.

Lorman, S. "Simply thinking you have insomnia might cause health problems" (24 Nov 2017) CNN.

Mansukhani, MP, et al. "Apps and fitness trackers that measure sleep: Are they useful?" (Jun 2017) 84(6) Cleveland Clinic Journal of Medicine—"In general, these devices have major shortcomings and limited utility, as they have not been thoroughly evaluated in clinical populations."

Meltzer, LJ, et al. "Comparison of a commercial accelerometer with polysomnography and actigraphy in children and adolescents" (1 Aug 2015) Sleep 38(8) 1323–1330—Commerical devices have "a significant risk of either overestimating or underestimating outcome data including total sleep time and sleep efficiency."

Morley, J, et al. "Digitalisation, energy and data demand: The impact of Internet traffic on overall and peak electricity consumption" (Apr 2018) 38 Energy Research & Social Science 128–137—"Peaks in data appear to fall later in the evening, reflecting the use of online entertainment."

Mortazavi, S, et al. "Blocking short-wavelength component of the visible light emitted by smartphones' screens improves human sleep quality" (1 Dec 2018) 8(4) Journal of Biomedical Physics and Engineering—Small study supports "hypothesis that blue light possibly suppresses the secretion of melatonin more than the longer wave-lengths of the visible light spectrum."

National Health Service. "How to get to sleep" (1[illegible] Jul 2016).

Palavets, T, et al. "Blue-blocking filters and digital eyestrain" (Jan 2019) 96(1) Optometry and Vision Science—"A filter that eliminated 99% of the emitted blue light was no more effective at reducing symptoms of DES than an equiluminant ND filter."

Paterson, JL, et al. "Sleep schedule regularity is associated with sleep duration in older Australian adults" (9 Oct 2018) 41(2) Clinical Gerontologist—"Sleep schedule regularity may be associated with sleep duration."

Perez Algorta, G, et al. "Blue blocking glasses worn at night in first year higher education students with sleep complaints: A feasibility study" (1 Nov 2018) 4 Pilot and Feasibility Studies.

Phillips, A, et al. "Irregular sleep/wake patterns are associated with poorer academic performance and delayed circadian and sleep/wake timing" (12 Jun 2017) 7 Scientific Reports—"Irregular sleep and light exposure patterns in college students are associated with delayed circadian rhythms and lower academic performance."

Price, C. "Putting down your phone may help you live longer" (24 Apr 2019) New York Times—"By raising levels of the stress-related hormone cortisol, our phone time may also be threatening our long-term health."

Robbins, R. "Sleep myths: An expert-led study to identify false beliefs about sleep that impinge upon population sleep health practices" (17 Apr 2019) 5(4) Sleep Health.

Schecter, A., et al. "Blocking nocturnal blue light for insomnia: A randomized controlled trial" (Jan 2018) 96 Journal of Psychiatric Research—Blocking slightly "improved sleep in individuals with insomnia symptoms."

Stillman, J. "Science has identified a new sleep disorder caused by sleep trackers" (4 Apr 2018) Inc.

Tanier, M. "Next big thing: Sleep science is becoming the NFL's secret weapon" (5 Oct 2016) Bleacher Report.

Van der Lely, S, et al. "Blue blocker glasses as a countermeasure for alerting effects of evening light-emitting diode screen exposure in male teenagers" (Jan 2015) 56(1) Journal of Adolescent Health—"BB [blue blocker] glasses may be useful in adolescents as a countermeasure for alerting effects induced by light exposure through LED screens and therefore potentially impede the negative effects modern lighting imposes on circadian physiology in the evening."

Xie, J, et al. "Evaluating the validity of current mainstream wearable devices in fitness tracking under various physical activities: Comparative study" (12 Apr 2018) 6(4) JMIR mHealth and uHealth—"Fitness trackers of different brands vary with regard to measurement of indicators and are all affected by the activity state, which indicates that manufacturers of fitness trackers need to improve their algorithms for different activity states."

Younes, M. "Technology of sleep monitoring . . . Consumer beware!" (27 Jun 2017) BioMed Central Network—"Consumer devices overestimate sleep time and miss approximately two-thirds of awake time . . . There is no convincing support for the claim that these devices distinguish light from deep sleep or REM from non-REM sleep."

The Relax, Dammit Rules

Broniatowski, DA, et al. "Weaponized health communication: Twitter bots and Russian trolls amplify the vaccine debate" (Oct 2018) 108(10) American Journal of Public Health—"Whereas bots that spread malware and unsolicited content disseminated antivaccine messages, Russian trolls promoted discord."

Diresta, R. "The complexity of simply searching for medical advice" (3 Jul 2018) Wired.

Edelman Trust Barometer, "Global Report" (2018).

Gallup, "Confidence in institutions" (2018).

Ghenai, A. "Health misinformation in search and social media" (2–5 Jul 2017) Proceedings of Digital Humanities Conference—"People can be potentially harmed by search engine results."

Holone, H. "The filter bubble and its effect on online personal health information" (Jun 2016) 57(3) Croatian Medicine Journal 298–301—"The algorithms that support us in finding relevant information quickly can also bring us closer to a gravitational black hole of information, which subsequently can lead us to make bad decisions about health issues."

Kahan, DM. "Why smart people are vulnerable to putting tribe before truth" (3 Dec 2018) Scientific American.

Kelly, J, et al. "This is what filter bubbles actually look like" (22 Aug 2018) MIT Technology Review.

Kiss, SJ, et al. "Balanced journalism amplifies minority positions: A case study of the newspaper coverage of a fluoridation plebiscite" (2018) 43(4) Canadian Journal of Communication 633–645.

Knight Foundation, "American views: Trust, media and democracy" (16 Jan 2018).

McNeil Jr, DG. "Russian trolls used vaccine debate to sow discord, study finds" (23 Aug 2018) New York Times.

Nicas, J. "Google has picked an answer for you—Too bad it's often wrong" (16 Nov 2017) Wall Street Journal.

Ortega, JL. "The presence of academic journals on Twitter and its relationship with dissemination (tweets) and research impact (citations)" (20 Nov 2017) 69(6) Aslib Journal of Information Management 674–687.

Russell, FM. "The new gatekeepers: An institutional-level view of Silicon Valley and the disruption of journalism" (2019) 20(5) Journalism Studies 631–648.

Scullin, M, et al. "The effects of bedtime writing on difficulty falling asleep: A polysomnographic study comparing to-do lists and completed activity lists" (Jan 2018) 147(1) Journal of Experimental Psychology 139–146—"To facilitate falling asleep, individuals may derive benefit from writing a very specific to-do list for 5 min at bedtime rather than journaling about completed activities."

Vosoughi, S, et al. "The spread of true and false news online" (9 Mar 2018) 359 Science 1146–1151—Research on social media has found that falsehoods "diffused significantly farther, faster, deeper, and more broadly than the truth," likely because lies are simply more interesting than the truth.

Welbers, K, et al. "Social media gatekeeping: An analysis of the gatekeeping influence of newspapers' public Facebook" (11 Jun 2018) 20 New Media and Society 4728–4747.

INDEX

N

O

P